HUXLEY AT WORK

Toine Stokes

Huxley at Work

With the Scientific Correspondence of T. H. Huxley and the Rev. Dr George Gordon of Birnie, near Elgin

MICHAEL COLLIE

Emeritus Professor of English at York University, Toronto

M
MACMILLAN
PRESS

First published 1991

Published by
MACMILLAN PROFESSIONAL AND ACADEMIC LTD
Houndmills, Basingstoke, Hampshire RG21 2XS
and London

Companies and representatives
throughout the world

Printed in Hong Kong

British Library Cataloguing in Publication Data
Collie, Michael
Huxley at work: with the scientific correspondence of T. H. Huxley and the Rev. Dr George Gordon of Birnie, near Elgin.
I. Title
560.92
ISBN 0–333–51059–3

Contents

Acknowledgements

I am most grateful for the assistance I have received during the writing of the book, from Jean Archibald, Special Collections, University of Edinburgh Library; Mr and Mrs Cowie, the Manse, Birnie; Jenny Hurst, Geological Survey, Keyworth; Dr David Iredale, Moray District Record Office; Colin McClaren, Archivist and Keeper of Manuscripts, University of Aberdeen; Mrs. Jacqueline Martin; Photographic Services of York University; Dr. Michael Benton; Wei Ping Lu; Ian Morrison, the Falconer Museum, Forres; Sinclair Ross; and Mrs June Mead.

My special thanks are due to Jennifer McLaughlin for her friendly encouragement during the whole period of the writing of the book; Christine Sangster for her unfailing assistance and guidance with the papers and specimens in the Elgin Museum, and for helpful reading of the final typescript; and Mr Ewan Gordon for allowing me to publish the Huxley and Gordon letters in his possession.

In a survey of well over a hundred research libraries in the United States and Britain I have been greatly helped by archivists, librarians and curators of manuscripts whose expert and courteous assistance it is now a pleasure to acknowledge.

List of Plates

Preface

A few years ago I became aware of that substantial archive of documents in the Elgin Museum that represents the life work of George Gordon, notably in the fields of geology, palaeontology, archaeology and local history. During the course of his long residence in Moray as Minister at Birnie, a few miles from Elgin, he corresponded with many of the leading scientists of the mid-Victorian period, including James Nicol, Sir John Lubbock, Darwin, Sir Roderick Murchison and T. H. Huxley. Usually Gordon retained drafts of the letters he wrote. Consequently the Elgin Museum contains, not only thirty four previously unpublished letters from Huxley to Gordon, but also drafts of twenty five letters from Gordon to Huxley in Gordon's hand.

The correspondence between Huxley and Gordon, which is not social but exclusively scientific, concerns the unexpected discovery of reptilian fossils and footprints on the south shore of the Moray Firth, and the step by step interpretation of them during the twenty year period between 1858 and 1878. Suddenly Huxley had in his hands the fossil remains of what turned out to be formerly unknown extinct reptiles – a marvellously challenging research opportunity. How could it be that in the remote district of Moray, so many hundreds of miles north of London, there had once existed reptiles of the kind Victorians associated with jungles and deserts? Was it really true that, where the town of Elgin now stood, there had once lived twelve foot long creatures with scales like those of a crocodile and long tails (*Stagonolepis*) and smaller, squatter creatures with beaks and powerful hind legs (*Hyperodapedon*), and that these were reptiles that became extinct before the era of the dinosaur? Had Britain really been so different then, more than two hundred million years ago, that land-based reptiles could flourish in it? For a while, there was turmoil and perplexity. In the School of Mines in Jermyn Street in London, Huxley was well-placed to conduct the research needed to decipher, analyse, describe and place the fossil remains of these previously unknown creatures. In Huxley's mind, this research, which for the most part he conducted at a distance, had a direct bearing on some of the questions which concerned him most, including the as yet ill-defined relationship of geology and palaeontology. This was partly because the established work habits of descriptive geologists had not previously been challenged by the quite independent findings of a palaeontologist. His correspondence with George Gordon on this subject therefore relates to one of the most important aspects of his professional career.

As soon as it had been decided that the letters discovered in Elgin were worthy of publication both because they had a direct bearing on the course of Huxley's career and on the development of palaeontology in Britain, it became important to know whether any other letters of Huxley's or Gordon's had survived elsewhere that either concerned reptilian fossils specifically or the Moray Firth generally. All the Scottish University libraries have been visited for this purpose, as have the National Library of Scotland, the National Trust of Scotland and the Scottish Record Office. In England, a thorough search was conducted in the libraries of institutions with which Huxley was associated: the British Library, the Royal Society of London, the Geological Society, the Royal Institution and elsewhere. Only in the libraries of the Geological Survey of Great Britain, at Keyworth, and Imperial College in London were other autograph letters discovered. In North America, enquiries have been made at all the libraries that are members of the American Association of Research Libraries, as well as other great collections throughout the United States. Although American libraries have rich holdings of autograph material relating to Huxley's scientific work, there is nothing that relates specifically either to George Gordon or to palaeontological research in Moray, except for a single letter in the library of the American Philosophical Society.

The correspondence being presented here is therefore self-contained. It tells the story of a great Victorian puzzle – were there really *reptilian* fossils in the Old Red Sandstone of the Moray basin? – while historically it also represents the steady progress towards scientific understanding that is illustrated by the publications of both men. For Huxley, the reconstruction of extinct reptiles from fossil fragments was a triumph in itself, the more so because it brought him to a view of evolution that incorporated evidence of the kind he respected.

In the organisation of this book there are two parts, which complement one another. It would have been possible to have utilised Huxley's letters in a narrative that absorbed excerpts from them in a conventional manner, but for several reasons this strategy was rejected.

In the first place, to present the correspondence intact, unaltered, verbatim in Part Two permits a clear focus on a geological and palaeontological puzzle that for mid-Victorians was unique to Elgin. It was a puzzle which eventually commanded the attention of all the principal geologists of the day, including Roderick Impey Murchison, Charles Lyell, Adam Sedgwick and Archibald Geikie. While the interests of most of these people were geological, Huxley entered the arena exclusively as a palaeontologist. Because the work necessarily proceeded by fits and starts, anyone reading this book needs to keep in mind that the Huxley–Gordon correspondence can be seen in retrospect to have concerned (at least for Huxley) only three sets of reptilian fossil remains:

those of *Stagonolepis, Telerpeton* and *Hyperodapedon*. As far as Elgin was concerned Huxley did not allow himself to be interested in anything other than these three creatures. The title *Huxley at Work* refers strictly to what he did towards the identification, description and analysis of the remains of these reptiles, and to the publications that resulted from his research. There is only one exception to this – his brief remarks on a fourth reptile which he called *Dasygnathus longidens*. But these brief remarks do not refer to actual research. Whether Huxley's work on fossil finds in the Elgin region might have been more extensive is difficult to determine. Suffice it to say for the moment that he concentrated exclusively on the three sets of fossil remains named above. His correspondence with Gordon concerns only these reptiles and, if letters passed between them on other subjects, they have not been preserved.

A second reason for the book's strategy concerns George Gordon, and Gordon's special relationship with Huxley. Gordon was a minister of the Church of Scotland, Huxley a convinced agnostic. When Huxley visited Moray (which he only did twice), he stayed in the Gordon Arms in Elgin, not at the Manse at Birnie. Nonetheless they managed to work together for a number of years in a reasonably harmonious manner. During the course of this co-operation, Gordon was by no means subservient to Huxley. He knew the region well, cared about it, and was prepared to spend many hours attempting to satisfy Huxley's research needs, as well as his own. Though mistaken on some important points, he stuck to the task of finding and interpreting fossil remains as best he could. He was a thoughtful, generous, energetic amateur scientist and the correspondence celebrates this.

There is a third reason for preserving the correspondence intact and for reproducing the letters in their entirety. Generally speaking the complete texts of Huxley's letters, whether social or scientific, important or unimportant, are not at present available either to the scholar or the general reader. Of course, many of the letters in the archive of Imperial College were used by Leonard Huxley in his biography, but it was no part of his purpose to assess the archive from a bibliographical or editorial point of view. Neither the accuracy of the transcriptions nor the scope of Leonard Huxley's approach have been seriously tested. Excerpts from different letters have appeared elsewhere, usually without any discussion of provenance or textual accuracy. There is no edition of Huxley's correspondence. The recent survey of research libraries has shown that there are many more autograph letters from Huxley in other collections than are listed in standard manuscript guides. It is difficult to see the point of further scholarly work on Huxley until these letters have been collected, edited and published, since they contain a mine of as yet unused information, while taking them into account, on the other hand, might challengingly modify our overall view of an important nineteenth-century scientist. It is thus in the context of Huxley studies as a whole that a proper editing of the Huxley–Gordon letters was preferred.

Before a word of this book was written, questions were quite reasonably asked about the appropriateness of the title – *Huxley at Work*. The title is intended to emphasise the fact that the book's purpose is to throw light on what Huxley actually did. In other words an analysis of how he conducted his work as a palaeontologist took precedence over an analysis of his publications, though these are discussed in some detail. His words are interesting enough, but can one get behind the words? The correspondence, messy and incomplete though it is, helps one to do this. No claim is being made for the intrinsic intellectual or literary merit of a single letter, though some are extremely interesting. Rather the letters, together with the annotation of them, are presented as a means of access to T. H. Huxley's mind. His mind did not work in quite the way that many people have assumed it worked.

Finally there is the question of terminology. Any writer of this type of book probably wants to satisfy both the specialist and the general reader. Such at least is the case here. For the sake of clarity, there has been some sacrifice of specialist terminology, with the attendant risk that some sections may seem terminologically naive to the expert. This risk will have been worth taking, however, if the general reader can see the book as a contribution to the history of nineteenth-century science, a contribution which both attempts to show an important figure in a clear light and to provide some indication of what early palaeontological work in Britain was actually like.

Huxley became a palaeontologist at a time when the discipline was in its infancy. The earliest *Oxford English Dictionary* definition of the term is from Charles Lyell's *Elements of Geology* (1838) where it is said to mean 'The study of extinct organized beings; that department of geology that treats of fossil animals and plants'. Huxley was not what is now often called a palaeobotanist; he never studied fossil plants. Nor was he a field geologist. In as far as this book, however, concerns the very beginning of the systematic study of vertebrate fossils in Britain, it is not too much to claim Huxley as one of the country's first palaeontologists. Incidentally, the spelling of the word seems of little consequence, though many may prefer the modern 'paleontology'. In this book, Huxley's own normal spelling of the word has been preferred.

In a wonderful old house called *Les Merceries* in the Charente, a house filled (until the recent regrettable death of its owner) with works of art of great originality, I once saw a life-size representation in wood (from a single block) of two figures in seventeenth-century French court dress, the woman's right hand just resting lightly on the man's left arm. There was space between them, but the genius of the sculptor had captured the closeness of their relationship. Thomas Hardy refers to this space at the beginning of *The Mayor of Casterbridge*. A man and woman are walking apart down a country lane, but by their manner it can be seen that, though they walk apart, they are man and wife. Someone once told me that, when she entered a bar to keep an appointment with two friends, a man and a woman, she perceived instantly, as she threaded through the crowd, that there existed an intimate relationship between them, even though their manner was formal and there was a decorous space between them. These three instances are metaphors for the intimate relationship that exists between Part I and Part II of *Huxley at Work*. Though they can be read separately, they are designed to remain together.

Part One

1

Huxley the Public Figure

For Thomas Henry Huxley (who will now be called Huxley throughout), the essential foundation for a scientific career was hard work. He subscribed to the work ethic of his peers. Something was bound to come of the effort, if one applied oneself in a disciplined way to a chosen task; work produced results and the compulsion to work needed no further justification. Inspiration and insight were all very well, but steady application was the way to get things done. If there were some fields of academic enquiry that could be advanced by speculative brilliance, there were others that yielded their truths to steady, disciplined, informed, day-by-day investigation, and Huxley vastly preferred the latter. The exercise of work-bench palaeontology suited his character. For this type of science to be effective, though, one had to know what other people were doing, keep in touch with them, weigh and respond to their ideas, read their papers and books – all of which took time and self-discipline. In fact, it took most of the time there was, and Huxley, particularly in his earlier professional years, seems to have begrudged the merely social hours that had nothing to do with science. Not for him the mercurial assimilation of ideas of the bar-room wit; he preferred to retire to his own study to work out a line of thought, limited though it might at first seem to be, that could not be faulted on either logical or factual grounds. He tended to be highly critical of anyone else's line of argument if it could be so faulted.

Huxley's scientific career made him a public figure. This seems to have been less a function of his success as a scientist, and more the result of the form his youthful ambition happened to take. He wanted to be a person of public consequence, seeing that this was one of the ways in which science could be advanced. It was also the means by which Huxley could be advanced. During the twenty year period covered by the present book – roughly 1858 to 1878 – he was *inter alia* a founder member of the Metaphysical Society, President of the British Association (1870), Rector of Aberdeen University, Secretary and afterwards President of the Royal Society, and Secretary, later President of the Geological Society. Wanting to be powerful, he was successful in achieving a public position in which he could not be ignored. He wanted to be more than one among many; he wanted to lead, dominate, be recognised as eminent, and his naturally competitive temperament initially permitted him to work towards these goals without his feeling that anything of a different kind was being sacrificed.

It has been demonstrated, most recently by Ruth Barton in the *British Journal for the History of Science*,[1] that he achieved and maintained an influential position partly, or maybe largely, through the agency of the X Club. This was a London dining club whose nine members – John Tyndall, Edward Frankland, George Busk, John Lubbock, Joseph Hooker, Herbert Spencer, William Spottiswoode, Thomas Hirst and Huxley himself – during the winter met each month prior to Royal Society meetings, thus constituting a kind of lobby group within that organisation. These friends cared a great deal about what they regarded as pure science (with engineering and technology excluded), believed on the basis of their personal experiences that science could only prosper if actively safeguarded from religion, and were prepared to engage in whatever political manoeuvering might be necessary, within the associations and societies of which they were members, to ensure that their new values would prevail against the old guard. The principal forum for this activity was the Royal Society, of which Huxley had been a Fellow since 1851. But even without the X Club Huxley would still have been the kind of career-oriented person who appreciated the advantages of a high profile. When he applied his mind to purely scientific problems he expected the results to be noticed, and he expected his public pronouncements to carry weight because of his scientific reputation.

Back in 1851 he had said in a letter that 'fighting and scratching to keep your place in the crowd excludes almost all other thoughts'.[2] Remarks of this kind are not easy to interpret, the more so, perhaps, because no satisfactory biography of Huxley has yet been written – that is, one of sufficient psychological depth. Self-justifying remarks sometimes have to be discounted, at least to a certain extent. Still, he was certainly very sensitive to the crowd, where 'the crowd' meant that bustling fraternity of scientists, intellectuals, savants – amateur and professional alike – that dominated mid-Victorian London. In this company, the company of like-minded people whom he respected, Huxley would be the bull-dog (as his popular reputation alleges him to have been) in any confrontation between science and religion. This was not simply because of his own agnosticism; it was also because he knew his peer group – the members of the X Club, for example – invariably approved of attacks and counter-attacks against the Church. On the other hand, when religion and metaphysics were not involved, he took great care not to offend those members of the crowd with whom he worked, sometimes appearing surprisingly, even unnecessarily diffident. He was instinctively a social being, someone whose career came first, but someone who preferred to further that career by getting on with people who mattered. It will be seen that this paradox of temperament, his being at one and the same time both intellectually aggressive and politically or socially cautious, did affect the way he worked. Nowhere is this more clear than in the twenty year spell of work he devoted to the palaeontology of the Elgin region.

Whatever may have been excluded from Huxley's life for the sake of professional advancement, involvement in any type of association or society meant active involvement. By 1858, when the story of the Elgin reptiles effectively begins, he had not only kept his place in the crowd, but was already a prominent, nationally-known scientist, a widely respected, wholly dependable person whose counsel was sought at the highest levels. During the period under review, he served on a number of important enquiries, including the onerous Fisheries Commission in Scotland and the Commission of Science and Art Instruction in Dublin. If he was ambitious, he was also dedicated, strongly believing both in science itself, and in the contribution he found he was able to make towards its progress. How else but by working hard could all this be achieved? An 1857 photograph by Maul & Polybank shows him in the self-assured pose of the successful, professional scientist, the public figure. This was a man whose physical appearance ensured that he would not be ignored. The photograph gives emphasis to the dark hair parted at the left and falling just over his collar at the back, the rather ferocious and bushy dark eyebrows, the prominent and well-formed nose and the large, sensual mouth. The frock-coat, with tweed waistcoat and trousers, and jaunty cravat, more hint of the man about town than of the adventurer, explorer or worker in the field. Of course he had dressed up to have his photograph taken. But these powerful features have also a hint of the sardonic, making one wonder how ruthless this human being might be if the need arose. There would be little scope for ruthlessness, one would suppose, in the study of the fossil remains of those extinct reptiles that are the subject of this book.

Of a house in North America where (at a later date) he had especially appreciated the warmth of his hosts' hospitality, he said approvingly that it was almost as quiet as his own home in London, a remark that hints both at the long hours of solitary labour that success involved and, as well, at the tiptoe existence to which other members of the household were sometimes confined while the great man worked. Not that this is extraordinary in itself; countless scholars have had to struggle to reconcile the intellectual and the domestic. Obviously Huxley read a great deal. What emerges from any study of even part of his long and distinguished career is the extent of his control over the 'literature' of any discipline that interested him. He and his friends made a point of knowing what was happening in France, Germany and North America; the proliferation of professional journals had made this possible; to attend meetings where people had read foreign books and articles that one had not read oneself was undesirable. Maintaining the pace certainly required energy – there was so much that had to be done – but Huxley was by daily habit the scholar, so much so that he was invariably able to place his experimental and workbench researches within the context of his wide reading. This fact turns out to be of importance in his teasing out the significance of the reptilian fossils found in the north of Scotland.

Huxley, though, despite the long hours in the study, was far from being a recluse. On the contrary, to whatever extent he may have worked into the small hours at home, during the day he kept himself visibly at the centre of scientific London through his Professorship at the Royal Institution, his Professorship at the School of Mines, his lectures at the Royal College of Surgeons, and his daily contacts with fellow scientists, including life-long friends like John Tyndall. Much personal business seems to have been transacted at his club, the Athenaeum, and at afternoon or evening professional meetings nearby, at Somerset House (the first home of the Geological Society) or in Albemarle Street (where the Royal Institution was and is situated). The places with which he was associated – the Royal Society, the Royal Institution, the School of Mines, the Geological Society, the Royal College of Surgeons and the Athenaeum – constituted his territory, his home ground, and because all these institutions were close to each other and could easily be reached on foot, he could easily be met. Huxley's position was different from that of many late-twentieth scholars, who so often have a longish list of institutions, institutes and libraries around the world that they need to visit without having the funds to do so. Because Huxley read other people's published work and heard them give talks, and because he also read and assessed their papers when they were submitted to the Royal Society or the Geological Society for publication, there was not really very much in this confined London world that escaped his attention.

As an essential or integral part of the story of Huxley's work on the reptilian fossils that were found near Elgin, it has to be remembered that, having neither telephone or fax machine, he wrote an immense number of letters, both as a means of 'home' or local communication (the penny post ensured same day delivery in London and, with any luck, next day delivery in places like Elgin) and of communication 'abroad' with those colleagues throughout the world on whose research findings he so much depended. The letter – given prominence in the present book – was an indispensable part of his working life, a vehicle of research enquiry and the lifeline with scientists outside London. How many hours each day he spent writing letters cannot easily be determined, but it must often have been several. Certainly, the incessant letter writing, without secretarial assistance, contributed to the overwork that later affected him in ways difficult to anticipate in his early years when he was full of energy and no task could be thought to present an insurmountable difficulty. But the network of communication was essential to the progress of his research. His correspondence with George Gordon amply illustrates this.

Some people nowadays find it difficult to accept the letter as a *normal* method of communication, either because, as is often argued by way of special pleading, the letter has altogether lost its utility in the fast-moving late-twentieth century (that is, when two-litre cars travel through London at the same speed as a hansom cab did in 1858) – 'nobody writes letters these days' – or because it has acquired an aura of excessive formality which prevents its being seen as an easy, vital, reliable, spontaneous, ordinary form of discourse. From a Victorian point of view, though, the possibility of communicating quickly by mail was an exciting advance, in the simple sense that things had become possible that hadn't been possible before. Thus the exchange of letters between Gordon and Huxley was not on the periphery of their research; it was central to it. Nothing would have happened at all had they not been able to converse through the Royal Mail.

While private letters were one of the means by which Huxley conducted his research, an important aspect of the high visibility Huxley wanted and achieved for himself was his academic writing and publishing, though only a fraction of it will be referred to in this book. The sheer volume of his published work was and is impressive. Between 1858 and 1870 he published twenty five articles in the *Quarterly Journal of the Geological Society* alone. Since he wasn't strictly speaking a geologist, the simple thing to say about this is that, because geology was the central, most influential, most generally accessible scientific discipline of the early Victorian period, it was natural for Huxley to wish to play a prominent role in the Geological Society, and to regard its journal as an important career medium. During the same period he made substantial contributions to the *Proceedings of the Royal Institution*, the *Proceedings of the Zoological Society*, the *Journal of Anatomical Physiology*, the *Transactions of the Ethnological Society*, and the *Philosophical Transactions* and *Proceedings* of the *Royal Society*. He published on average what nowadays would be called eight major papers a year, a remarkable rate of production if sustained, as it was. Furthermore, as his career developed, he responded generously to editors' requests for articles, as was consistent with his conviction that a scientist had a responsibility to make his ideas known to the general public, and with his evident love of putting pen to paper, like any other writer who just enjoys writing. Impossible, therefore, not to know him, or know of him.

With this pattern of unremitting labour went the stress of controversy. Well beyond the call of duty, he found it difficult to restrain himself from championing the cause of science even when it was attacked by the transparently prejudiced or ignorant. 'I fought the parsons at Aberdeen and got beaten but the press was on my side and their victory will not serve them long'.[3] Of course, George Gordon, on whom Huxley at first completely depended for his palaeontological work in Scotland, was – if not exactly a parson – nonetheless a Minister of the Church of Scotland, something which from an X Club point of view was just as bad. What happened when Huxley and Gordon met at the British Association meeting in Aberdeen, a dozen years or so after they had begun working together? Was Gordon one of the parsons against whom Huxley fought? It must have been by some internal accommodation that he managed to deal in a civil manner with a Minister of the Church whose opinions he would otherwise have detested. Looking back from a time when the Christian church is so uninfluential, in Britain at least, that people of other persuasions do not have to take up arms against it, Huxley's habit of entering into the lists against the clergy looks like a waste of energy. Why not spend that same energy doing better work or writing better articles? Not only was his tendency to be drawn into controversy taxing in itself, it also resulted in some repression of Huxley's feelings as regards belief. Perhaps he protested too much.

An example of this is the argument he later had with Theodore Powys which Powys recorded in the following terms as an 'extract from a long discussion with Huxley', and which shows, perhaps, how obstinate Huxley could be.

> You have been educated as a Christian gentleman; and although you have ceased to believe in Christ and God, you will from force of habit still adhere to the Christian code of conduct. The injury which you agnostics are doing will not come in your own time, but afterwards. Up to now, whatever the mistakes made by Churches and ecclesiastics, Christianity has created a code of justice, mercy, humility, valour, and virtue of which the British Empire has been a great example. Contrast our Empire with the ancient pagan empires and the difference speaks for itself. Christian chivalry, Christian power of sympathy, has been a civilizing agent which may yet be perfected, but its reversal will lead back to chaos and savagery. We Christians try to do our duty to God, Sovereign and Country; and we endure adversity bravely on earth in virtue of our hope of heaven. What have you to give to the younger generation in exchange for what you take away?
>
> Huxley looked at his watch and said he was late for another engagement.[4]

In his opposition to religion, his distrust of metaphysics and his intolerance of Christian conviction, Huxley seems to have believed that space had to be negotiated for science and that it wasn't enough simply to get on with one's own work, letting both the science itself and its implications speak for themselves. Many of the contemporary ecclesiastics were themselves aggressive when it came to controversy about science and religion. Science was more of a threat to them than religion was to the scientist. Too much was at stake for them to ignore all that geology and palaeontology asserted and implied. So Huxley was repeatedly drawn into disputes which a hundred and more years later seem unnecessary and unproductive. All credit to him, therefore, (in one sense) for the restraint he showed in his dealings with the Rev. George Gordon.

Indeed it would be quite wrong to suggest simplistically that, for the sake of his public career, Huxley was callous in his dealings with others. When fighting and scratching resulted in illness and breakdown, people overlooked his pugnacity and rallied to his support, for the very reason that he was a highly respected public figure, and one who had materially affected the lives of many thinking people by liberating them from what they took to be superstition and prejudice. If they did not know the technical, detailed, professional papers he contributed to the Geological Society, they had access to his ideas through his published lectures, and more popular articles in the *Westminster Review*, the *Fortnightly Review* and, from 1870, the *Contemporary Review*.[5] By being articulate and knowledgeable himself, he literally taught them to think in a new way, so by any standard he was a force to be reckoned with.

Among all these people, there were many who found helpful his direct challenge to church authority. They liked his clear-headedness, sure logic and directness of approach, as well as his refusal to put up with cant and pseudo-religious humbug. His high literacy allowed easy access to his ideas – especially to the scientific basis of his agnosticism. He was influential, therefore, among those whose active minds needed freedoms that Church and State seemed not to provide. In this regard, it was his fate to be admired by people whom he never knew. When the first serious breakdown of health occurred in 1872, an unknown person approached him in the Athenaeum and offered to put 'his purse at his disposal', an offer that was declined at the time, but accepted later when the same individual left Huxley a substantial sum of money in his will.[6] The precise motives of the anonymous donor cannot be determined, though they must have stemmed from admiration and respect. At that time, Huxley probably did not understand what, in his case, was the relationship of thought and work. That was to come. As will be seen, a group of eminent scientists, at a still later date, raised £2100 to let him take a rest from work. Evidently they could perceive the damaging results of unremitting labour in ways which he could not, or chose to ignore.

The correspondence between Huxley and George Gordon recently brought to light in the Elgin Museum illustrates, in the field of palaeontology, *how* Huxley worked. This emphasis on the *how* is part of what James Secord in a recent review called the 'welcome trend towards the study of the material circumstances of scientific enquiry'.[7] The emphasis will be on what he actually did, how he thought, how he pushed his researches forward. And this account will be partly descriptive, partly analytical. The correspondence neatly complements a set of articles that he wrote during the same time span, articles that are listed in Appendix I. By looking at the two side by side one can see how problems were formulated and imagined, and also what lines of thought had priority in Huxley's distinctive way of doing things. Readers of *Huxley at Work* will understand that the articles and correspondence taken together represent only a small, though important part of Huxley's total output, so that the description, discussion and analysis of them has to be conducted with caution. They let one into Huxley's mind, but perhaps not the whole of it. Nonetheless, taken together they do at least show how he tackled a set of problems that were important to him.

The question of how the history of science might best be written is nowadays much debated, so it may be necessary to indicate at the outset that an assessment of Huxley's palaeontology in late twentieth century terms is not the primary purpose of the present book, though it cannot and need not be avoided completely. The historical method preferred here involves the attempt to see why Huxley's work was important at the time it was conducted. If, as well as reconstituting a set of mid-nineteenth century problems and solutions, one can also recover the immediacy of intellectual puzzles, the excitement of discovery and the pleasures of understanding, so much the better. The question is – why *at that time*, in the middle of the nineteenth century – were problems of palaeontology in the Moray region particularly worthy of attention? Why were they important to Huxley? Why did he devote so much time to them?

To answer these, and associated questions, we need first to establish what the difficulty was, after which the ways in which Huxley and George Gordon tackled it can be described by reference to their correspondence.

These opening remarks have not been solely designed to provide a general background against which Huxley's work on the Elgin reptiles can best be seen. Rather, the purpose has been to identify from the beginning some of the factors that affected the way in which Huxley conducted his scientific enquiries, at least as far as palaeontology was concerned. Had someone written a letter of recommendation on Huxley's behalf in 1857 or 1858, that person would have known about Huxley's ambition, but perhaps would have called it determination, or commitment to science. The psychological hazards of the work ethic would have been discounted, the erudition praised. The writer would not have called Huxley a palaeontologist, because the problems of the Elgin fossils had not yet been properly taken up (though a first article – 'On the Method of Palaeontology' – had appeared in 1856). Probably he would have called him a physiologist, a zoologist or even a naturalist and, if he knew him well, would have noted that great store was placed on logic, method, systems, classifications and anatomical structures and organisations, as well as work. Huxley was of the new generation of British scientist, that is, not an amateur like George Gordon who pursued problems for the love of the enquiry, however inconclusively, but a public figure who believed he had a duty to demonstrate that society needed the new knowledge that science could provide. Some recent work in the history of science has tended to adopt a sociological perspective – the scientist as professional, the scientist as imperialist, and so on – but my emphasis here will be more psychological than sociological. Of course the two are not mutually exlusive. Huxley cared about his social position, liked the power that went with professional success, benefitted from imperial technology, and seems not to have questioned ideas of national, racial or gender superiority. While being a child of his time in these respects could hardly be described as culpable, since he was only doing well what other professional people in mid-Victorian Britain also wanted to do, fearing the alternatives, or not regarding them as attractive, there is nonetheless no doubt that one could usefully analyse Huxley's total career from a sociological point of view.

Were that to be done, not only would his professional success have to be described, examined and assessed as a function of his social position, but, as well, his rhetoric would have to be taken seriously as an instrument of class power. No prose is more open to deconstruction than his. What is being argued here, however, is that, while Huxley's social sense furthered his work, various psychological factors inhibited it, placing limitations on the way he worked, and making him disinclined to examine freely a number of issues he knew to be important. The same analysis of his discourse that would give access to class imperatives would reveal a mind held in check by inner fears that had to do only with him and were not shared, or were not shared to the same extent, by friends like John Tyndall and Joseph Hooker. He was a psychologically repressed mid-Victorian whose negative mindset determined the way he worked, something one can say descriptively without detracting from the distinction of so much that he did. The kind of person Huxley was and the kind of work he produced were closely related. What will be shown as an illustration of this relationship is a research enquiry in progress, and the question that will frequently arise is – why did the participants think and behave as they did?

2
The Land of the Cakes

Huxley was not a geologist, though for a long period an active member of the Geological Society in London, and for a number of years its President. There are no pictures of him with rucksack and hammer investigating cliffs and quarries. Nor has any evidence survived of his accompanying friends and colleagues on field trips, even occasionally.

One of the puzzles of the Elgin story has to do with Huxley's willingness to extract from all the activity going on around him two or three specific problems of identification. How did he isolate the topics he believed were worth his attention? He knew about all the geological survey work that was being done, or had been done: in the South-West, for example, by de la Beche; in Wales by Sedgwick and Murchison; in Scotland by MacCullough, then Sedgwick and Murchison again. He well knew that this was the time of serious survey work, both because of the urgent interest in coal and mineral exploration (thus the School of Mines), and because of the challenge of the subject itself. Roderick Impey Murchison was a senior colleague, but, colleague or not, it would have been impossible for Huxley to remain ignorant of Murchison's pioneering survey work in the 1820s in the Hebrides, Caithness, Ross and Cromarty, Moray and Aberdeenshire. Huxley, under the shadow of a great man – Murchison – seems to have perceived that if he wished to make a significant contribution to this continuous bustle of activity, it would have to be as a palaeontologist and, if this is correct, it could be called a coincidence or a lucky chance that he just happened to be presented with an excellent academic opportunity in the form of fossils found in the North of Scotland, near the town of Elgin, in an area geologists still did not understand. As far as that area was concerned, the exploration and mapping by the Geological Survey of Scotland only got under way after Huxley had done the main part of his work. That work was done, therefore, during a period of uncertainty.

Sometimes it happens that a person who wants to understand other people's activities will go to observe them. Nineteenth-century novelists, for example, began to instruct themselves on aspects of Victorian life that ordinarily would have been outside their realm of knowledge: hospitals, slums, brothels, mills, ironworks, prisons. In the same investigative, if detached spirit one might go down a coal shaft, or join one of the parties that went to see Brunel's tunnelling under the Thames. Victorian Britain was after all an information society. They preferred to know, even if, on some fronts, they were then prepared not to talk about what they had learned. It is puzzling that Huxley kept any tendency to geologise strictly in check. He never, or hardly ever, went into the field.

For early Victorians, field trips consisted either of local antiquarian explorations, often by clergymen who knew their district well but not any other (George Gordon and his clerical friends in the North of Scotland were in this category), or of vast, frequently pedestrian journeys in the course of which considerable tracts of land were surveyed, old maps corrected and the data for new maps established. Huxley had no ambition, it would seem, to emulate the men who made such journeys: Buckland, de la Beche, Lyell, Sedgwick, Murchison and many others. He was so much a town intellectual, so much a person of the closet, that to imagine him tramping about Wales, for instance, enduring the daily discomforts and disruptions of survey work, is virtually impossible. It is not even clear that he shared his contemporaries' interest in and love of landscape; his horizons were in his head.

On the other hand, he knew personally all the prominent geologists of his day, and he attended those early, pre-professional, animated, sometimes even rowdy meetings of the Geological Society at which most things of importance were routinely reported. He learned from them. But he himself would only tackle work that had not been tackled before. For career purposes, it was convenient that Scotland was a long way away. The Elgin fossils could be his.

He became President of the Geological Society because of his erudition and because of his intellectual eminence, not because of his field work. This meant that he was always cautious when dealing with men who *were* geologists and could speak from experience. He was willing to write about the theory of geology and to lecture about its importance, but strictly speaking he was not a geologist at all but a palaeontologist.

One of the consequences, however, of Huxley's close contact with men who were geologists was that the boundaries between the two disciplines were perfectly clear. He would not claim for himself an expertise outside the range of his own scientific activity, considerable though that was. He looked to others for the geological knowledge that his palaeontology required, reading widely, and regularly consulting colleagues like Charles Lyell and, later, Archibald Geikie. He was in close contact, too, with the staff of what was at first called the Geological Survey of Great Britain, though even here, because it was his habit to isolate a problem in order to think about it clearly, it is likely that he concentrated mostly on the information that was relevant to his immediate purposes. Some scholars are stimulated by their colleagues' talk in labs, common rooms, cafés, private homes and so on, to extend their own research enquiries, perhaps at the same time modifying their concepts in accord with the extension of interest. Huxley was not like this. He was an intellectual who limited consultation to what he had predetermined was strictly relevant. Concepts like 'Scotland' and 'Morayshire' were too amorphous to be helpful to him, which meant, ironically, that the need for consultation sometimes became quite severe. Years after he had begun his work on the reptilian remains found round about Elgin, he still wasn't sure where exactly they had been found. When Geikie sent him an off-print of an article on what Huxley called the 'Scotch Drift', he said in his letter of acknowledgement: 'I want to know where my reptiles are'.[8] He still didn't know and had little enthusiasm for going to see. London had been, and always remained, his place of work, and even as far as his Elgin work was concerned, he for the most part preferred to stay in the south, despite being inconvenienced by the lack of topographical and geologic maps.

This is not to say that Huxley did not know Scotland at all. When he told Geikie in a letter that he was 'off to the land of the cakes' (meaning, of course, oat cakes), it was to a meeting of the Herring Commission in Aberdeen. Various types of work took him to Scotland fairly frequently. During the period covered by this book he attended all but one of the meetings of the British Association that were held in Scotland. He lectured in Edinburgh, made many friends there and was later given an honorary degree by Edinburgh University. His son was an undergraduate at St. Andrews, where he himself took at least one holiday. As Rector of the Universities of Aberdeen, he visited the city fairly frequently and the papers in the John Webster collection attest to the extent of his involvement in University business. He even claimed for himself some knowledge of the Scottish character. Exclaiming 'My experiences of Edinburgh enable me now to understand the peculiar harmonies and vigour of the Scottish temperament', he dryly explained to Professor Blackie that this temperament was the result of a slow evolutionary process. 'All the weak and susceptible are destroyed in early youth by your summers'.[9] He had, however, mostly a book and tourist's knowledge of Scotland, being more familiar with its railway lines than its lines of fracture and fault.

In his book, *The Making of Geology – Earth Sciences in Britain 1660–1815*, Roy Porter usefully demonstrates that geology was not a nineteenth-century invention, a reminder it would obviously be foolish to forget. Geological enquiry already had a certain momentum by the time the great nineteenth-century geologists set to work, though the early focus of that enquiry was as often as not theological, having more to do with theories of creation in a Biblical context than with scientific research, analysis and interpretation. The antiquarian spirit of eighteenth-century

Britain, indeed of Europe as a whole, had nonetheless made people think about what they saw with their own eyes, for example sea-shells on mountain slopes, and various religious, philosophical or pseudo-philosophical theories about the beginning of life on earth. If the world had been created quickly, why had God put sea shells on mountains? Or was there a case to be made for successive acts of divine creation? There was also the religious problem of fossils that provided evidence of extinct forms of life. Why had God created a species only to render it extinct? At the same time, say, that Huxley was elected to the Royal Society, these and similar questions could only be naively formulated because the great achievements of nineteenth-century geology had yet to make their impact.

Of the Scottish School of Geology, as he calls it, Roy Porter said: 'The second defining feature of Scottish Earth Science [the first being a lack of ideas] was an *enduring* concern, even into the nineteenth century to realise it as "the natural history of the mineral kingdom" – "within a philosophy of natural history"'.[10] Whether accurate in absolute terms or just a way of expressing something, this sort of view might have been acceptable in the Manse at Birnie, where natural history did not challenge belief, but it would have had no interest whatsoever for Huxley. Geology in itself, as he understood it, provided sufficient reason for not taking literally the Christian Book of Genesis. From his point of view, eighteenth-century geology, whether English or Scottish, was old-fashioned geology, as he himself was from time to time disposed to explain in papers and talks that reviewed the development of the subject since the Greeks. Modern geology was for him strictly an affair of the nineteenth century. Its basis was stratigraphy – the study of observably different strata of rock lying on top of each other. Its principal textbook was Charles Lyell's *The Principles of Geology*.

Despite his medical education, but (as will be seen) in ways entirely consistent with it, Huxley had soon developed an interest in 'the testimony of the rocks' (to use the title of Hugh Miller's famous book), an interest which in retrospect can be seen to have satisfied him because of the apparent directness of the correlation between evidence, if judged accurately, and conclusions derived from that evidence, if derived rigorously. Rocks told the truth. Fossils also told the truth, if they could be properly analysed. There may have been continuity in the earth sciences over a longish period, but Huxley was almost exclusively interested in the new methodology and the concepts that went with it. For Huxley to be a practising scientist, however, rather than someone wholly dependent on other people's investigations, he needed experiments of his own. Scotland provided the opportunity for this, though at arm's length. In a sense, Elgin became his laboratory, the place where, at that time, he alone worked. It therefore did not matter to him if others continued to be interested in natural philosophy. If pondering God's purposes was part of the earth sciences tradition in Scotland, then Huxley made a break with that tradition.

For the purposes of the present presentation and assessment of the scientific correspondence between Huxley and George Gordon, it has to be remembered that in nineteenth-century Britain the means available for analysing, plotting and generally understanding rock strata were limited. One had for the most part to work with what could be seen on the surface. The sea concealed much of what one wanted to know; the limit was low tide. On the land, instruments for drilling deep bore holes did not exist. Coal shafts were important, but at that time not many coal shafts were deep. Shafts could not be sunk merely to satisfy the curiosity of the geologist. Furthermore, instruments for exact mineralogical analysis did not exist either. The mineral nature of rocks could be observed and recorded, but the mineral content itself could not be dated with precision. Consequently, speculation about the age of strata had to be based upon the strata themselves or their fossil content.

When Huxley began to think about the fossil fragments that Gordon started to send him, the correlation between the strata in which they were found and the fossils themselves could not easily be established. Everything had to be done from scratch, and without the equipment and methodology that would later be taken for granted. The rock had not been identified. The fossils were unique. This is why Huxley spoke about the need to reconcile geology and palaeontology. Even though geologists had used fossil evidence for many years to confirm the identification of rock strata, Huxley was unwilling to state in print that vertebrate fossils found in a certain type of rock would *only* be found in that type. He did not know where they might yet be found. Logic would protect this negative position until new positive evidence presented itself. It was not logical to claim that, because a reptilian fossil had been found in Triassic rock, it might not be found in Devonian. This characteristic conservatism in research (or non-research) inclined him to bide his time until new 'facts' came his way – as it will be seen in the correspondence he believed they eventually did. In other words, to understand Huxley's predicament when he started this particular bit of work, one has to take into account all the things that were then not known. To do this makes Huxley's contribution to the early development of palaeontology in Britain more clear.

Huxley as palaeontologist was for a while at odds with some of his geological friends because, understandably, one of the features of early stratigraphical research was an interest more in the *normal* disposition of layers of rock than in exceptions to the norm. This particularly applied to the study of the Old Red Sandstone round the Moray Firth, but it also applied elsewhere. The first need was to observe the continuities of the system. Only in a second stage of work could one begin to make sense of observed disruptions of that system. To understand the story being told in this book one has to know that the disposition of rock around Elgin did not conform to any norm, but that Huxley and his colleagues did not yet know this. It was more reasonable, at the beginning, to anticipate continuity, than unconformity.

In the vast enterprise of geological survey – the study of stratigraphy in the whole of the British Isles, in the whole of the world even – it was an article of faith that the ascending or descending order observed in one place would normally be repeated in other places. Carboniferous would always have Devonian beneath it; inconceivable that the order would be reversed – anywhere. Permian would always be beneath Triassic, and so on. This being the case – that is, this being a reliable form of new knowledge – the nineteenth-century geologist was perfectly prepared to argue by analogy. If, in a part of the world that had been explored and surveyed, rock of one type had a different type of rock beneath it, the discovery of the first type elsewhere necessarily established that rock of the second type would be beneath it, even if that fact could not yet be observed. It was on this line of thought that mid-Victorian geology was based, though in 1858 everyone was very conscious that a huge amount of work still had to be done, so much so that you could still make a satisfactory career for yourself as a descriptive stratigrapher, without bothering overmuch either about exceptions to the norm, or about the causes of those exceptions.

Huxley himself propounded this principle, the argument by analogy, in the following terms.

> Allied with geology, palaeontology has established two laws of inestimable importance: the first, that one and the same area of the earth's surface has been successively occupied by very different kinds of living beings; the second, that the order of succession established in one locality holds good, approximately, in all.
>
> The first of these laws is universal and irreversible; the second is an induction from a vast number of observations, though it may be possible, and even probable, to have to admit of exceptions. As a consequence of the second law, it follows that a peculiar relation frequently subsists between series of strata, containing organic remains, in different localities. The series resemble one another, not only in virtue of a general resemblance of the organic remains of the two, but also in virtue of a resemblance in the order and character of the serial succession in each. There is a resemblance of arrangement; so that the separate terms of each series, as well as the whole series, exhibit a correspondence.[11]

Huxley spoke these words at the anniversary meeting of the Geological Society during the period in which he was corresponding with George Gordon, and while it was still unclear (that is, had not yet been demonstrated) whether or not the geology of the Moray region conformed to the principles Huxley adumbrated in his talk. He believed it had to. But he had colleagues who did not.

At all events, what the early nineteenth-century geologist felt he had to do, as geologist, was to produce maps that recorded his knowledge of stratigraphy. The maps of Smith, MacCullough and Murchison were early representatives of this genre. One only has to examine, for example, the map of Wales in Murchison's *The Silurian System* to appreciate what a massive achievement it was, given the early date. How was geological work of this kind done? It could only be done by a close, on the spot observation of rock strata as revealed in outcrops, fissures, river valleys, railway cuttings, quarries and cliffs, and in visible unconformities where one type of rock could be seen to be in contact with a different type. Not everything *could* be seen. One therefore had to gather evidence as best one could, without the help of the sort of technology taken for granted in the twentieth century, recording the evidence in drawings, maps and sections, and subsequently trusting those drawings, maps and sections as the *only* record of what had been seen until, perhaps decades later, a new spurt of work permitted new interpretations. Murchison's description of the north of Britain for a while enjoyed this privileged position. Huxley could not challenge that position without irrefutable evidence.

The correct interpretation of stratigraphical evidence around the town of Elgin could not easily be achieved. Few people knew the region well. Early, partial or inaccurate accounts of it had to stand until someone took the trouble to re-examine the evidence on which they were based. That meant more field work. In the case of Morayshire this took the greater part of the century. No wonder people were confused. And no wonder that the letters of Huxley and Gordon bear the marks of this confusion. Even as late as 1887, on the evening when Huxley (to anticipate) gave his final paper on the *Hyperodapedon Gordoni*, Archibald Geikie 'recalled the early controversies about the age of the Elgin sandstones, and pointed out that no satisfactory stratigraphical solution to the problem had yet been found'.[12] Geikie, the pupil, was being loyal to the master, Huxley, when he said this. By contrast, the President of the Geological Society gave a commonsense reply to Geikie, in effect asserting what could only be the solution. Nonetheless, the conversation on that occasion was a reminder of how long the Elgin sandstones had puzzled them all, and how difficult it had been for geologists and palaeontologists to come to agreement with each other.

It is time to be clear about the specific nature of the puzzle. Early studies of the Old Red Sandstone in the area of the Moray Firth by people like Hugh Miller, Adam Sedgwick and Roderick Impey Murchison had not sufficiently pondered the fact of two major strata of rock not being represented in the region at all i.e. the Carboniferous and most of the Permian. (See Plate 1.) Had they done so, they would have been forced to ask themselves questions about the geologic history of the region. What events or series of events, so many millions of years earlier, had resulted in there *not* being these two massive sediments? How, specifically, had the region taken the form they found it had? These were the larger questions of the kind Huxley left to geologists. There was also a second set of questions, closer to home. Preoccupied by normal sequence and trusting in analogy, they did not sufficiently think about, to the extent that they noticed them at all, the region's distinctive features. They failed to identify the unconformity separating the Old Red Sandstone, containing fish remains of Devonian age, and the New Red Sandstone, containing reptilian fauna of late Permian and Triassic age. This was because of the complexity of the outcrops, and because they expected the descriptions of Miller, Sedgwick and Murchison to be accurate. The towns of Elgin, Burghead, Lossiemouth and the like were built of stone quarried locally. It was in these quarries that quarrymen found reptilian remains, and their counterpart moulds, as well as footprints. It was thought that reptilian remains had been found in the Old Red Sandstone. Had this been the case they would have been the earliest reptiles then known. Local folk, including George Gordon and his friends, were understandably excited by the possibility that their region was both unique in itself, and also uniquely interesting in an important modern way. They had been taught about the continuities of the Old Red Sandstone in Caithness, Cromarty and Moray, which indeed they believed they could see with their own eyes. Had these continuities not been frequently described? Had everyone so far been wrong?

Here, for example, is how the region is described in the 1851 edition of Anderson's *Guide to the Highlands*, the edition closest to the commencement of Huxley's work on the Elgin fossils.

> Morayshire contains neither Triassic, Permian nor Carboniferous rocks; but those of the Old Red Sandstone are very well displayed, and several of the strata abound in icthyic remains, although as yet no trace of Molusca or Crustacea has been found. As is generally the case with this series, the classification of its members, from their included fossils, does not correspond with that of any other district. Many of the beds are unfossiliferous, so that a rigid definition of the limits of the divisions is impracticable. The uppermost of these consists of grey, yellow and red sandstones and conglomerates, both fine and

> coarse, associated in some places with chocolate-coloured shale; there are also occasional deposits of ore or less siliceous limestone. The ridges of Stotfield, Covesea, Inverugie and Roseille, belong to this division. Its thickness is considerable, but notwithstanding diligent search, it has only produced a single fossil, the *Stagonolepis Robertsoni Ag*, found at Stotfield by Mr. Duff.

Patrick Duff was a friend of Gordon's and a major figure in the palaeontological exploration of the whole region. *Stagonolepis Robertsoni* became the subject of Huxley's first paper on the problems of the north east of Scotland.

The account continues without a break:

> The second division is composed of sandstones and siliceous conglomerates of various hues, and sometimes containing calcareous matter; seams of chocolate-coloured shale and fuller's earth; limestones like those above them, and at Cothall, on the Findhorn, above the limestone, a green clay with calcareous nodules. The strata of Quarrywood, and the moors of Cardon and Alves, of the magnificent section on the Findhorn, of Scat Craig, and of the Lossie and Shoggle at Birnie, are included in this division. The limestone beds are unfossiliferous; but the other strata generally yield other osseus relics of fishes, or the impressions of them in greater or lesser abundance.

There follows a list of fossils described by Agassiz in his *Monographie des Poissons du Vieux Grès Rouge*, a book acknowledged as one of Huxley's starting points. Then the tantalising sentences:

> There have been found, besides these, many species as yet unedited. The conglomerate of Scat Craig abounds in fossils, and many are also to be extracted from the rocks of the Findhorn. Beautifully perfect impressions of scales and osseus plates have been discovered in the Bishopsmill and Hospital quarries, and in those of Carden moor.[13]

Here exactly is the problem. 'Scales' and 'osseous plates'? Of a fish? Until Huxley began his work it had seemed reasonable to think so.

It is unlikely that anyone who visited Moray in the mid-1850s with this guide in hand would have had any reason to doubt its accuracy; indeed the long geological note, obviously written by an insider, had an authoritative ring to it. Having been told not to look for the Triassic, he would like not see it. Only a particularly alert traveller would have been perplexed by the references to scales and osseus plates. The writer's claim, a little further on, that the whole area must have been disturbed by 'powerful subterranean forces', would have to stand until someone came up with a better explanation of its unconformities. All in all, this passage, contained as it was in what the author called a 'synopsis or index of the subject' (actually a footnote excerpted from the longer account by Alexander Robertson that had appeared in earlier editions) had all the marks of academic reliability. Visitors did rely upon it until, that is, scholars far away in London began to ponder some of the inconsistencies in the received account that it recorded.

As soon as academic and professional geologists were alerted to the puzzle in the claim that in the north-east of Scotland reptiliferous fossils had been found in the wrong, or at least extremely unlikely, strata, two obvious questions arose: first, had the geology of Moray been correctly described? Second, what exactly were the supposed reptiles the quarrymen had found? Huxley was far more interested in the second than in the first of these questions.

In Mario di Gregorio's *Huxley's Place in Natural Science* there is a useful section about the way in which Huxley began his palaeontological work. Given that his analysis and description of *Stagonolepis* was one of his earliest forays into serious palaeontology, how did he learn how to proceed? In the first instance he was helped by a colleague in the School of Mines, J. W. Salter, whom di Gregorio describes as an 'entomologist and palaeontologist of considerable repute'.[14] Huxley would also have read various papers that had appeared in the early 1850s including G. A. Mantell's 'Description of the *Telerpeton Elginense*, a Fossil Reptile Recently Discovered in the Old Red Sandstone of Moray' (1852) and Brickenden's 'Notice of the Discovery of Reptilian Foot Tracks and Remains in the Old Red or Devonian Strata of Moray' (1852), but these would not have taught him what to do in practice; nor would they have established the principles governing the analysis of fossil remains. The first occasion on which he applied his chisel to a piece of rock must have been momentous. No doubt it is correct to say that in early laboratory experiments with the chisel Huxley was helped by Salter. No doubt it is also correct to note that his thinking must have been profoundly influenced by Agassiz, though the two men did not know each other very well. (A separate new study is required of the relationship of Huxley and Agassiz.)[15] But because of Huxley's training in anatomy while a medical student, and his preoccupation with classification based on structural organisation, it will be convenient here to go back to Georges Cuvier to establish how Huxley himself worked. He knew Cuvier's works in French, including his *Recherches sur les ossements des quadrupèdes*, but it will be easiest for present purposes to refer to an early English edition of one of his books, that is, Robert Kemp's translation entitled *Essay on the Theory of the Earth* (1813), where a revolution in scientific thinking is identified in clear terms.

'Naturalists', says Cuvier in *Essay on the Theory of the Earth*,[16] 'seem to have scarcely any idea of the propriety of investigating facts before they construct their systems'. He says that 'the cause of this strange state of affairs may be discovered by considering that all geologists have hitherto either been mere cabinet naturalists, who had themselves hardly paid any attention to the structure of mountains, or mere mineralogists, who had not studied in sufficient detail the innumerable diversity of animals, and the almost infinite complication of their various parts and organs'.

After these strictures, Cuvier goes on to establish what he considers to be a set of basic principles, using quadrupeds as his example, but in an argument of universal applicability. First:

> The appearance of their bones in strata, and still more their entire carcasses, clearly establishes that the bed in which they are found must have been previously laid dry, or at least that dry land must have existed in its immediate neighbourhood. Their disappearance as certainly announces that this stratum must have been inundated, or that dry land had ceased to exist in that state. It is from them, therefore, that we learn with perfect certainty the important fact of the repeated irruptions of the sea upon the land, which the extraneous fossils and other productions of marine origin could not themselves have proved; and by careful examination of them, we may hope to ascertain the number and the epochs of these irruptions of the sea.

In the pages which follow, Cuvier reviews the reasons for believing that 'none of the larger species of quadrupeds, whose remains are now found embedded in regular rocky strata, are at all similar to any of the known living species'.

What Cuvier says here about quadrupeds informs Huxley's attitude to reptiles. Huxley's ideas about evolution in the 1850s were extremely tentative. They remained tentative through the 1860s, even after the publication of Darwin's *On the Origin of the Species*. But when he began his work on the Elgin fossils, he did believe in a creature's adaptation to environment by dint of structural or anatomical specialisation. If the environment did not alter, there would be no need for any living creature to alter. In fact, though, he found it difficult to think about alterations to structure. Once a type was established, might it not persist unaltered for a long time? How could a skeleton of a known type be modified? This for many years was a sticking point. In the study of fossils, Huxley began by believing that the task was to establish their taxonomy in terms of precise anatomical description, a description which was independent of, but which would logically lead to thoughts about how the creature, say a reptile, had functioned and survived. To distinguish between species within the class, *Reptilia*, for example, Huxley kept strictly to the task of identifying his fossils' anatomical features, his credo being that a correct observation of the anatomy would in turn permit correct classification – classification in this sense being, not an arbitrary ordering for pragmatic reasons, but a reliable kind of knowledge. Cuvier, of course, had not developed the dynamic sense of evolution that Darwin was to supply. In following Cuvier, Huxley was not only acquiring a research method that he could use in his study of the Elgin fossils, but was also confining himself to the relatively safe territory of taxonomic classification, in which large or heady speculation about geological causes or events was not called for, and where the discussion of the relevance of extinct reptiles to theories of evolution could be postponed, if not positively evaded.

In passing, it needs also to be said that Huxley only came slowly to questions of global distribution. If the relationship of anatomy to habitat could be established, then it was natural to expect, globally, that similar fossil remains would be found wherever the same geological conditions occurred. During the first part of the twenty year period being reviewed in this book, Huxley lacked evidence to think usefully about the global distribution of extinct species. The matters about which he corresponded with Gordon had to be settled first. The work of settling them, however, attracted attention, so that before long people began to send him specimens from other parts of the world.

One of the practical problems Huxley faced was how to study *incomplete* fossil remains. Obviously in many cases the skeletal remains of a reptile might have been scattered over a widish area after death. Alternatively, they might be scattered by quarrymen during the breaking up of the rock. Different parts of the skeleton might have been subjected to different types of pressure over the centuries, leaving them distorted and difficult to identify. What can the palaeontologist do when only fragments have been recovered? On this subject Cuvier said:

> Fortunately, comparative anatomy, when thoroughly understood, enables us to surmount all these difficulties (of reconstructing the whole from the parts) as a careful application of its principles instructs us in the correspondence and dissimilarity of the forms of organized bodies of different kinds, by which each may be rigorously ascertained, from almost every fragment of its various parts.

In other words, if you have parts of the fossilised remains of an animal, fish or reptile, even a very small part, or its counterpart (that is, either the fossil itself or its mould), you *can* reconstruct the whole.

This applies most obviously when the investigator is totally familiar with the anatomy, or what Huxley calls the organisation (nowadays perhaps 'body plan' or 'design') of the creature being reconstructed. Cuvier continued this line of thought in uncompromising terms:

> By thus employing the method of observation, when theory is no longer able to direct our views, we procure astonishing results. The smallest fragment of bone, even the most apparently insignificant apophysis, possesses a fixed and determinate character, relative to the class, order, genus, and species of the animal to which it belonged; insomuch that, when we find merely the extremity of a well-preserved bone, we are able, by careful examination, assisted by analogy and exact comparison, to determine the species to which it once belonged, as certainly as if we had the entire animal before us.

What Cuvier recommends in these paragraphs is what Huxley did. He applied the principles enunciated by Cuvier to the Elgin fossils, of which no other examples had at that time (1858–59) yet been found. He refrained from immediately going to Elgin to see whether or not his geologist friends had accurately described the terrain. Instead, he remained in London studying the fossil fragments George Gordon sent him. His initial boldness – that is, his complete confidence in the methods he adopted – was fully vindicated when he realised that large conclusions could indeed be drawn from fragmentary evidence. Even when he was not completely right, his pioneering work was of vast importance. He had found a legitimate subject: as yet unidentified fossils in the land of the cakes. To that subject he applied the best method he knew of at the time: Cuvier's. No wonder people paid attention to his first papers on palaeontological problems in Scotland.

It will be seen that he *did* eventually acknowledge the inconvenience of not knowing where 'his fossils' had been found. Staying in London, however, allowed him to implement the Cuvier method of comparative anatomy, at first in the strictest, most disciplined way imaginable in as far as no other examples of these reptiles were known, and a little later with even greater confidence when the reptilian remains found near Elgin could be compared with fossils found in other parts of the world. Admittedly a number of years were to pass before he could make even limited sense of what was sent to him, and admittedly his interpretations were not always accurate, as will be seen. Nonetheless, from his point of view he must have felt that he was on the right path; he was tackling a serious problem solely in its own terms by means of a method of enquiry that was wholly logical and practical. To begin to understand the material universe you had to ask the right questions in the right way. This is what Huxley was doing, and it seems reasonable to guess that he derived considerable satisfaction from a research enquiry that was completely self-contained. But was it self-contained? His data came from Elgin and someone else gathered it. Before saying more about Huxley's palaeontology, we need to look briefly at the life of George Gordon, the Scottish naturalist on whom Huxley had to depend.

3
The Intellectual Life of Elgin

In letter 13 of *A Tour Through the Whole Island of Great Britain,* Daniel Defoe remembered that the monks at an early date had claimed Elgin as their city:

> the Cathedral shows, by its ruins, that it was a place of great magnificence. Nor must it be wondered at, if in so pleasant a place, so rich, and so agreeable a part of the country, all the rest being so differing from it, the clergy should seat themselves in a proportioned number, seeing we must do the justice to say, that if there is any place richer and more fruitful, and pleasant than another, they seldom fail to find it out.

Defoe called Elgin a 'very agreeable place to live in' despite the fact of its being 'above 450 measured miles from London'. As for the town itself, he said:

> so here are a great many rich inhabitants, and of the town of Elgin in particular; for the gentlemen, as if this were Edinburgh, or the court, for this part of the island, leave their Highland habitations in the winter and come and live here for the diversion of the place and plenty of provisions: and there is, on this account, a great variety of gentlemen for society, and that of all parties and opinions.[17]

This town is some miles east of Inverness and a few miles south of the Moray Firth. Though few visitors to the region will be unaware of the majestic river Spey nearby, and the whisky and salmon associated with it, Elgin itself is on the river Lossie, which joins the Firth at Lossiemouth, to the north.

All the reptilian fossils referred to in this book were found in an area bounded by the river Nairn to the west and the river Spey to the east. There was discussion of fossil finds in the rocks along the north shore of the Moray Firth, but Huxley did not work on these. When the reptiles were alive, two hundred and fifty million years ago, the climatic and physical conditions around what is now the Firth must obviously have been completely different; different, that is, from the conditions that Defoe recognised, which are much the same as those anyone would experience today. It will be convenient to refer to Elgin as the centre of this territory, particularly because the fossil remains discovered during the nineteenth century are still popularly referred to as the Elgin fossils. The recently refurbished and beautifully managed Elgin Museum has always been the centre of operations for anyone interested in this subject, and there is now as excellent a display as the space permits of some of the important fossils brought to the Museum in Huxley and Gordon's time from nearby quarries.

Some two or three miles south of Elgin is the oldest church in Scotland that is still in use, at Birnie, which is the name, not of a compact, easy to locate village, but of an extensive, sparsely populated parish which now also takes in Pluscarden.

Early maps, such as the one reproduced in the 1851 edition of Anderson's *Guide to the Highlands and Islands of Scotland,* show that the old road south from Elgin passed by Birnie on its way to Rothes, the present A95 not then being in existence. The roads marked on this map, which is one of the most likely maps that visitors in the mid-nineteenth century would have used, are a reminder of how different the region then was. The main road from Elgin to Forres and Nairn followed the route of the A96 but many of the paved roads that would be used to explore the area nowadays, such as that between Lossiemouth and Burghead, scarcely existed except as unpaved farm tracks. In 1850 Spynie Loch had not yet been drained. Gordonstoun School had not been founded. The railway had reached Aberdeen but not Elgin or Inverness. The little railway from Elgin to Lossiemouth, which was to play such an important part in the Huxley-Gordon story, had not yet opened, though it was only a year or two away. Much traffic, indeed, was still by sea. Travellers went by coaster to Aberdeen from Burghead or Lossiemouth, and by the same route packages could be sent to London. Even as late as 1873, Huxley travelled back on the 'City of London' from Aberdeen to Wapping Old Stairs. Travel inland was on foot or by horse and trap. There is a photograph on display in the Elgin Museum showing geologists setting off in an open trap for a day's exploration – and this, much later in the century.

The sparsely populated District of Moray has to be imagined in quite other terms than those of today, with prominence given – topographically speaking – to gentlemen's houses and estates, to farms and harbours, and to the churches, such as the one at Birnie. Who would know this territory better than a Minister of the Church of Scotland who lived at the centre of it? Yet it has to be said, too, that during the period covered by this book (1858–78) both the Penny Post and the ever expanding railway system acted as liberating forces, greatly improving communications within Britain, and permitting a type of collaboration between Gordon and Huxley that would have been much more difficult at the beginning of the century. At least for palaeontologists, the movement of specimens was as important as the movement of ideas.

Nor was Elgin quite as Defoe had known it, though still benefiting from its temperate climate. Though the Kirk was still strong, as were the traditions for which it stood, thoughtful people realised that being in the Provinces, and being church goers by long tradition, in no way prevented their coming to terms with the modern knowledge they acquired or knew about from their reading and travels. On the one side was religion: the (for many) comforting daily or weekly habits of religious practice; the values associated with religious belief – patriarchy, duty, obedience, work, gender inequality; and the psychological benefits and deficits of an interior religious life – a sense of purpose, which might inspire a lifetime, counterbalanced by a sense of guilt, which might contaminate a lifetime. On the other side was an awareness of change and progress, of colonial expansion with its new opportunities for the inhabitants of the glens, and of equivalent escape routes for the intellect, like the one that seemed to be offered by Huxley's agnosticism and belief in science as an agent of social reform. Like many small towns of comparable size, Elgin was a microcosm of the unresolved conflicts of mid-Victorian life. In times of war, the young men would join the Gordon Highlanders, patriotically, as though established values were secure. In time of peace, at least some part of the population devoted itself to mental improvement, whether for recreation, like George Gordon, or for some more earnest purpose. There was a tradition of intellectual self-improvement thereabouts. What was talked about in Edinburgh and London could also be talked about in Elgin. George Gordon and his friends ensured that this was so.

One of these friends was Patrick Duff, the town clerk (1791–1861), a modest, intelligent man with the leisure and inclination to think about the place in which he found himself. He carried forward the discussion of the geology of Moray some fifteen years after the visits of Murchison and Sedgwick, which had been in the 1820s. Thinking about the familiar landscape, not in terms of the use to which it had traditionally been put, but of the geological forces that had shaped it, was itself an intellectual revolution, but one which Duff and his associates could cope with. They had all read their Hugh Miller, not least the highly influential book *The Old Red Sandstone*. It was in what he and his friends understood to be a new mode of thought that Patrick Duff had written *Sketch of the Geology of Moray*, which first appeared as a series of articles in the *Elgin Courant* (resulting in their being widely discussed) and then as a book published in 1842 by Forsyth & Young in Elgin, Oliver & Boyd in Edinburgh, and Smith, Elder in London. The book probably represents pretty accurately how Patrick Duff, George Gordon and their friends thought about one of their common interests – geology – during the 1840s.

Because Patrick Duff had no conception of an ice age, he was unable to imagine the effects on the terrain of moving ice, or its weight, or the consequences of its beginning to melt, but he did know that the vast volume of sand in the Elgin region, for example at Culbin, needed to be explained. Besides this, in order to make sense of the inland raised beaches where he and his friends were finding 'extensive beds of shells identical with those pervading the shore at the present time', he speaks of 'subterranean forces, such as are at present exhibited in the earthquake and the volcano' and of 'electric and magnetic influences'. He understood that some great force had been exerted at some point of time, but he did not know what that force might have been. Patrick Duff represented the enlightened part of Elgin life, the part open to new ideas and actively reconsidering its environment in the light of those ideas. Interestingly, in his concluding chapter, he states:

> Every person entering on the study of geology should fix in his mind, and keep steadily in view that all rocks, with the exception of the volcanic, which are only an altered condition of the original deposits, have been deposited under water, either that of the sea or inland lakes; that all hills have been caused by the upheavement of these deposits by subterranean forces in the form of volcanoes or earthquakes; that all valleys and hollows have been occasioned by the abrasions of torrents and rivers, and not infrequently by the sinking down or subsidence of portion's of the earth's crust into the vacuum caused by the elevation of the other parts. This last conclusion is supported by the appearances presented everywhere by rocky strata; for scarcely a cliff of any extent occurs that does not show several of these appearances, technically called faults, which are occasioned by portions of the mass sinking down for want of support below, while the great bulk remained at the elevation to which it had been forced up.[18]

In the history of ideas or the history of science, which is more important – the adoption of wrong or inadequate conclusions, or the identification of the right problems? And in the history of geology, which deserves most attention, the general idea – such as the idea of successive catastrophes to explain geological change – or the example of the person at work who wants his conclusions to be based on what he himself has observed? Though this, if pursued, might turn out to be a false opposition, Duff's *Sketch of the Geology of Moray* provides an example of the latter, while at the same time it probably indicates the extent to which a powerfully Presbyterian community was able to accommodate new, potentially subversive ideas.

Duff's fourth chapter – 'Old Red Sandstone or Devonian System' – is also instructive, in as far as it provides a good illustration of how local folk thought about the Old Red before the discovery of reptiliferous fossils. Duff's book contained three colour plates. The first is an amateur geological map of the territory between the Spey and the Nairn. The names of the sites which later produced reptilian fossils are not given; the fossils had not yet been found.

Interestingly, though, the notes to the non-colour Plate V, drawn by Duff's brother-in-law, identify three reptiliferous fossil fragments as 'Fossils of the Wealden': a vertebra of 'Plesiosaurus Subconcavus'; an unclassified fossil 'supposed to be one of the bones of a reptile'; and some teeth of 'Plesiosaurus'. Of the last, Duff remarks: 'The largest specimen had a root, and had been inserted into a deep socket in the jaw, as in all reptiles, the teeth of fishes being fixed on the flat surface of the maxillary bone'. These were hints, but hints whose significance was not appreciated until other fossils had been found. The question of the implantation of teeth was to become important in the Huxley–Gordon correspondence.

The second colour plate in Duff's book is a section showing the 'Rocks present in Morayshire'. Of importance are the rocks *not* showing in Morayshire, and the right hand column gives a clear indication of the lack of knowledge that prevailed until after Huxley began his work. The bands represent geological periods that are too vast to be of much help in the dating of the complex deposits around Elgin.

This book of Duff's, together with a small number of essays by others, such as the one by Dr Malcomson that is referred to in the correspondence, set the stage. They represent an honest, local endeavour to understand in modern terms what was there to be observed. They powerfully influenced George Gordon. Modest as he was, he became preoccupied with the type of geology Duff had written about, becoming at the same time one of the leaders of the intellectual community that had at its centre the Elgin Museum.

The Rev. George Gordon (1801–93) who was inducted as Minister of Birnie in 1832, lived in the Manse across the field from the church, itself a fine, substantial building with stabling, outbuildings and various eighteenth- and nineteenth-century additions – not to mention the coal shed in which, on at least one occasion, Huxley and Gordon spent a few hours examining rocks. A special feature is the splendid garden, still tended lovingly by the present incumbent, Gordon Cowie, and in the nineteenth century the site of some of George Gordon's research as an amateur naturalist. Was there a dispute in *The Zoologist* about whether mice used their mouths to carry their young? Gordon had *seen* this happen when part of the garden was flooded. Could bees be studied scientifically? Gordon corresponded with Sir John Lubbock on the subject and had built for him specially constructed hives to facilitate scientific enquiry, including one built to look like a human habitation. Did blackbirds have only one brood each summer? He had observed in the garden a promiscuous blackbird having three broods and twenty-six offspring in a single summer. Let it not be said that anything occurring in Birnie escaped his attention. Unless his parishioners sometimes did.

The records of Birnie Church, stretching back for centuries were found, some of them, by Gordon Cowie a few years ago in a black box in the attic, shortly after he resigned his professorship in law at Glasgow. The papers inside the box, including the account of the region which Gordon wrote for the *New Statistical Account*, give an insight into parish life at Birnie which would be well worth further study by anyone interested in local history. An orthodox God-fearing, Bible-centred life was taken for granted; that heads of families would bring everyone to church on Sundays was also taken for granted; rural theology was probably not greatly threatened by eccentric local geologists. When the Church Elders met, it was not to consider the collision of science and religion, but more often than not to deliberate on the sexual habits and preferences of the younger parishioners. Not only blackbirds were promiscuous in nineteenth-century Birnie.

On 2 November 1851, for example, the Kirk Session met at the Manse to consider the case of one Elizabeth Robertson who, appearing voluntarily (so the records claim), acknowledged that 'she had been guilty of the sin of fornication and that Roderick McKenzie had been her partner in guilt'. This was typical. Something like 80 per cent of the nineteenth-century Kirk records concerned fornication, the arbiter of sexual behaviour being, not Darwin yet, but John Knox still. If only Elizabeth Robertson had lived a few decades later, the impact of the study of extinct fossils on the social *mores* of the day would have liberated her from the need to feel guilty about sex. She could have ignored the Kirk. She could have been negligent about marriage. She could have had Roderick McKenzie without having to explain herself to a group of old men in the Manse.

Whether George Gordon found other people's sexual activities absorbing is open to question. He thought the habits of extinct reptiles were more worth thinking about than the sexual proclivities of members of his congregation in places like the nearby village of Dallas where Elizabeth Robertson lived. When Elders of the Kirk of Birnie died, he sometimes neglected to replace them, and was twice visited by church authorities because of complaints on this score. Members of his congregation grumbled that he had taken to reading aloud from the Bible during Sunday services, instead of treating them to the sermon which their long journey to church made them feel they deserved. They could read the Bible by themselves at home, they said. Whether George Gordon ever used the pulpit to instruct the farmers of Strathspey on the advances in palaeontology learned journals told him were being made in Edinburgh and London is not known. The heads of families in Presbyterian Moray had a view of sexuality, and of human behaviour generally, somewhat different from what was implied by the failure of some species to survive. Had they been told that a species might be defined by its breeding habits, they might have nodded assent. But had they been told that monogamy was not an essential part of the definition, they would have been nonplussed. Gordon must have found ways of not upsetting them over much. For that matter, he must have found ways of not upsetting himself.

Although a conflict between the demands of science and the demands of religion may have been seen by some to have affected the Minister, most remembered the Manse at Birnie with simple affection, and one wrote:

> None who has ever visited the Manse at Birnie during his long incumbency is likely to forget the sweet spot, at once so cosy and elegant; the shrubbery with its rare shrubs, the garden with its wealth of 'fulgent' roses; the bees with all the apparatus for studying their habits, the fossils and arrowheads in the little study, and the drawing room presided over by a kind and hospitable lady.[19]

The Manse was George Gordon's base for a more thorough exploration of the surrounding countryside than had perhaps ever been conducted previously. Better than any man, he came to know what fish were in the Firth, what birds, insects and butterflies could be seen in the glens, moors and coastal district, at what sites there were ancient remains, and what plants had grown, traditionally, in its temperate climate. For his career as naturalist, as shown in part by the publications listed in Appendix II, Gordon was awarded an honorary degree at Marischal College in the University of Aberdeen.

Gordon was also responsible for creating the circumstances in which thinking people in and about Elgin could meet each other, in order to talk both about science, and about whatever other aspects of modern knowledge might interest them. In 1836, he helped found the Elgin & Morayshire Literary and Scientific Association, modelled on similar societies throughout the British Isles, and the Society (now the Moray Society) in turn built the Museum, which was opened in 1843 and in which local artefacts were collected. These included many of the fossils that were soon to be the centre of national attention, though, of course, local enthusiasts also kept some of their 'finds' in their own possession, as is noted in the correspondence reproduced in Part Two. For almost fifty years, Gordon in one way or another worked for the Society and for the Museum in the company of a fine band of enthusiasts.

When George Gordon retired, if that is the right word for such an active person, he was highly lauded for everything he had done for Elgin during the fifty years of his ministry at Birnie. To the speeches made in his honour he replied in the following terms:

> I must however attribute most of what Mr. Duff has said, and much of the hearty manner in which you have received what he has said of me, to the feelings of private friendship rather than to any deservings of my own. To the inscription of this epergne, you have been pleased to link my name with the words 'The Elgin Museum and Science'. It is only in the very subordinate walks of science that I have done anything of service – namely, in Natural History and Archaeology. What I have done for the Museum has been altogether a labour of love. I have done it for the pleasure of doing it, as a recreation to me, a healthy employment of the body and mind. Or, if I had any ulterior motive in view, it was to work out and make known the natural history and archaeology of the fine old Province of Moray, in which our lot has been cast, extending from the Beauly to the Spey, and from the nearest watershed of the Grampians to the Moray Firth.[20]

There is a singular appropriateness in the fact that the letters between Huxley and Gordon, which are being presented here in Part Two, should now be lodged in the very Museum to whose founding and early development he contributed so much.

Long before Gordon retired, however, the Museum achieved modest fame by being the place to which anything discovered locally of geological or palaeontological importance would logically be brought. The correspondence shows that the Directors of the Museum were scrupulous, even jealous guardians of the collection, anxious about the raids that Huxley made on it, but also properly proud of being the custodians of material that the best minds of the day felt constrained to think about, analyse and interpret. Eventually, after Gordon and Huxley had done their work together, the fame of the Museum gave rise to meetings such as the summer congress of 1881 when the Elgin Literary and Scientific Association played host to the Inverness Field Club, the Banff Field Club, the Ross-shire Philosophical Society, the Nairn Literary Association, the Aberdeen Philosophical Society and the Edinburgh Geological Society. It must have been a marvellous gathering – *was* a marvellous gathering according to the *Elgin Courant & Courier* for 2 August 1881. Elgin's reputation derived in large measure both from George Gordon's efforts over a long period, and the work he had done with the, by 1881, famous T. H. Huxley.

To go back to the beginning of the story, when Huxley, at the School of Mines in Jermyn Street, began to appreciate the significance of the reptilian remains in Elgin, what better guide and counsellor could he have found than George Gordon, a man of goodwill who already knew the region like the back of his hand and, more than that, already knew personally all the individuals who had made the discoveries. On what map in London would Huxley have found places like Findrassie, Spynie, the Knock of Alves, Scaat Craig, Stotfield and Covesea?

As a matter of fact, some of these sites are difficult to find nowadays, even with a good map in hand. The cliffs between Lossiemouth and Burghead have been further eroded. In the towns many of the sites have disappeared under streets constructed since Gordon's time. Quarries reported as sixty feet in depth when they were being worked are now much more shallow, or are filled with water. Trees have been planted around some of these sites, while others have become naturally overgrown. Exposed rock faces similar to those that produced fossils can be inspected, but the originals cannot. To attempt to visit all the sites today is, therefore, an unrewarding experience. To have attempted to visit them when the quarries were being worked would have been possible, but difficult without a guide. Because Gordon knew it all so well, Huxley soon discovered that he needed him.

Huxley needed Gordon, but George Gordon also had a mind of his own. The correspondence celebrates the meeting of two quite different intellects; the one convinced that the techniques of comparative anatomy would allow him to identify and classify in London the extinct reptiles whose fossil remains were found around Elgin, and the other convinced that it is better for a palaeontologist to know the region in which his fossils are discovered. Ironically, it turned out that Gordon's ideas about the geology of the region were largely mistaken, whereas Huxley, before he had ever visited Moray, arrived at a correct idea of the stratigraphical implications of the discovery of reptilian fossils there. At all events, they in effect worked together for a decade or more, the correspondence being the testimony of an exciting, certainly important, if in the final analysis less than satisfactory scientific partnership.

4

The Hours of the Chisel

The preliminaries over, it is now time to say something about Huxley's work on the Elgin fossils, from the time Roderick Impey Murchison first told him about them, up to the publication of *Memoir 3* of the Geological Survey in 1877.

Although he said in 1859 that 'the full description and illustration of all the remains which have been discovered will be reserved for the Memoirs of the Survey' (no doubt in compliance with a suggestion of Murchison's),[21] it was impossible to anticipate that the task would take him eighteen years. As already acknowledged, this presentation of his correspondence with Gordon isolates these years, artificially separating them from his career as a whole in order to focus on his palaeontology, and in particular the important set of articles that represented 'work in progress' on this subject.

The strategy of this section of Part I will be to follow the development of his thought through what he himself wrote, notice being taken both of the great care with which he proceeded, and of the need for frequent emendation and correction. The most directly relevant papers or presentations are listed in Appendix III. Several of them will be referred to in detail in this section. Of importance is the deliberate manner in which Huxley took account of new fossil finds, in what became a continuous process of thought and research leading to the major Memoir of 1877. Though always busy with other things (too busy), he refers to the countless hours he spent in the School of Mines with chisel in hand, attempting to tease out of the rock the secrets of the fossils that had been sent to him. To watch him at work gives us an insight into the type of scientist he chose to be.

As far as the Elgin fossils were concerned, Huxley had a set of objectives that can easily be stated. He wanted to establish that they were indeed the fossils of reptiles. He wanted to respond to the challenge of reconstructing an extinct fossil on the basis of minimal evidence. And he wanted to describe and classify whatever the fossils turned out to be in a manner of which Cuvier would have approved. He was helped in this, so he believed, by his extensive, and growing knowledge of crocodiles. Inevitably, these enquiries lead him to think about two other questions of great importance. The first was the question of global distribution.

Would the same reptiles be found anywhere in the world where the same conditions prevailed? The second was the question of extinction. Why had species that for a long period had been successful died out?

What follows is divided into three parts, each part, roughly speaking, being devoted to one of the three extinct reptiles in which Huxley interested himself, that is to say, the *Stagonolepis*, the *Telerpeton* and the *Hyperodapedon*.

The Stagonolepsis

His first Elgin paper was 'On the *Stagonolepis Robertsoni (Agassiz)* of the Elgin Sandstone; and on the Recently Discovered Footmarks in the Sandstones of Cummingstone'. The partial remains of the *Stagonolepis*, a creature previously entirely unknown anywhere in the world, had been found in a quarry at Findrassie, which is less than two miles north-west of the Elgin Museum. At one time a deep quarry much used for the construction of new houses in Elgin, it is now one third of its original depth and surrounded by Forestry Commission plantations. Earlier investigators, most notably Agassiz, had thought that these fragments were part of a fish. Agassiz *expected* to find fish fossils in the Old Red Sandstone. Huxley demonstrated that what had been found were remains of a reptile, in the process, therefore, initiating a controversy about what finding reptilian fossils in the supposed Old Red Sandstone meant.

It is important to notice Huxley's initial boldness. Although there had been hints of there being reptilian fossils in Elgin for some time, the reputation of the great palaeontologist, Louis Agassiz, was secure. Many had commented on the speed with which Agassiz made sense of fossil fragments where others had been defeated. There had really been no reason to doubt his original analysis and description of what Huxley now called *Stagonolepis Robertsoni* as a fish, especially as what he had analysed had been found in rock, so everyone believed, where fish fossils would normally be found. Though one hesitates to use the word 'intuitive' for a thought process that stretched out over several years – the years during the eighteen fifties *before* Gordon sent new evidence up to London – it is as though Huxley had an intuitive awareness of a problem that was big enough to be worthy of his attention.

Huxley's commitment to the resolution of the problem – to determine that the Elgin fossils were of reptiles not of fish, to reconstitute those reptiles, and to classify them – may also have been influenced by Owen's pronouncements about crocodiles. This is the Richard Owen who, as Hunterian Professor and Conservator at the Royal College of Surgeons, had already begun to publish his massive, four-volume *History of British Fossil Reptiles*. Owen had said that there were no crocodiles or alligators, or any reptilean creatures in Europe, and it was logical that there were not, since the right conditions were nowhere to be found. Might Huxley be able to show his arch-rival mistaken on this point?

He begins by telling the story of how the problem was first brought to his attention:

> On examining the bony remains associated with the scutes of *Stagonolepis*, some of which were preserved in the Elgin Museum and in the collections of Mr. Patrick Duff and the Rev. Mr. Gordon, while others were collected by himself, Sir R. I. Murchison was so impressed by their obviously reptilian characters that he used every exertion to gather together all the evidence which could tend to elucidate so important a question. In pursuing this object, Sir Roderick was aided in the most zealous and liberal manner by the Committee of Management of the Elgin Museum, by Mr. Patrick Duff, and by the very active personal exertions in the field of the Rev. George Gordon.
>
> To these two latter gentlemen my own thanks are also especially due for their prompt courtesy in attending to many inquiries and requests with which I have had to trouble them, since it became my duty, in accordance with the instructions of the Director-General of the Geological Survey, to enter upon investigation of these remains.[22]

The body of this first article is devoted to the analysis of the dermal scutes, ventral scutes and bones of *Stagonolepis* – that is, of all the bits and pieces he had in hand at the time. Huxley used the word 'scute' – defined in the O.E.D. as 'a large or bony plate, forming part of the integument of certain animals or fish' – to mean a rigid scale (see Plate 5) of the kind now readily associated with the armour of crocodiles and alligators.

For the techniques of comparative anatomy to work, the examination of these remains had to be minute and the measurements precise. In all that follows, it has to be remembered that from the measurement of one part of a skeleton (the part he had seen as a fossil or cast) he was making deductions about other parts (the parts he had not seen). Imprecision of measurement would therefore significantly increase the likelihood of error.

Huxley's prose style also had to be adequate for the task of communicating descriptions only partly supported by illustrations. From the time of Lord Bacon, and notably in Sprat's *History of the Royal Society* (1662), it had been understood that a scientist's style had to be plain, undecorated, economical and explicit. Huxley was not a mathematician; he probably trusted words more than he trusted figures. He moved from figures to words, believing that the former were the private, the latter the public mode of communication. Had he known about computers, and computer modelling, he would likely have been entranced, since design by extrapolation was what interested him. As it was, he lived in the world of 'correct' verbal usage, in which words had fixed meanings for educated people, and could be trusted, and in which an article or a printed book was necessarily one of the principal, and at the same time one of the most reliable methods of making a record of research and sharing it with others. Clarity of exposition was therefore as important as precise measurement.

Huxley's descriptive method, however, made such demands on the reader that one is glad now to be living in the age of the photograph. Thus he says, for example: 'The *irregular angulated scutes* (fig. 3) are pentagonal or rhomboidal, the ridge by which they are marked externally projecting so far backwards that their posterior margin (fig. 6) becomes triangular. One of the largest of these had a length of 1⅞ inch by a breadth of half an inch, and had a roughly pentagonal form, its anterior edges being slightly convex. Another had a length of 1½ inch by a breadth of ⅝th of an inch, and presented only a very small sculptured surface close to its anterior margin'.[23] One needs to have a first class visual imagination to be able to follow this kind of description. The plain, economical and precise prose style was an integral part of a method which required the listener or reader to be equally attentive and disciplined, if what was being described was to be visualised, although admittedly one had to remember, too, that some of the specimens could be exhibited when the paper was being delivered.

The written discourse of a Huxley paper had other distinctive features. In this first breakthrough paper, and also in the others that will be referred to in this section, he retains his preference for a straightforward linear narrative, reporting on what he has actually done and, by this means, allowing his listeners or readers to have the sensation of following the same path of enquiry that he followed himself. Whereas a paper nowadays might begin with a statement of intent, or be preceded by an abstract or synopsis, so that the conclusion at which the argument or demonstration aims is known to the reader at the outset, Huxley's method was not to anticipate the conclusion until the evidence had been presented. Obviously, this was artificial – a way of doing something that might have been done differently. Because Huxley is concentrating on the presentation of evidence in a narrative account the main line of which is a set of statements about what he actually did, he can allow himself to be guarded about the conclusions that might be drawn from his facts, thus giving the impression of great scrupulousness. Facts without ideas are like blackcurrant crumble without blackcurrants, but Huxley was preoccupied with what he called fact.

His method of presentation did not completely exclude, however, the type of interpretation represented by the two paragraphs from the *Stagonolepis* article that now follow:

At this stage of the inquiry, the full meaning of a piece of evidence, whose value I had, up till then, but very imperfectly recognised, became obvious. This was a remarkable natural cast, obtained by Mr. Patrick Duff at Findrassie, and which had been sawn through longitudinally by that gentlemen's direction, so as to expose its internal conformation. At first sight this curious fossil resembled nothing so much as the crushed and distorted cast of an *Orthoceras*, but both Mr. Patrick Duff and Sir Roderick Murchison had suspicions of its real nature, and in fact it turns out to be one of the most singular organic remains ever discovered, consisting of a natural cast of both the dermal bones and the vertebrae of a considerable segment of the tail.

Loaded by its heavy dermal plates, this caudal fragment appears to have sunk into fine siliceous mud, the accumulation of which had given rise to the Findrassie sandstone, and to have been completely permeated therewith, all the cavities left vacant by the putrefaction of the soft parts becoming filled up with a substance which soon hardened into stone. After this had taken place, the bony matter was, by some agency or other (probably the percolation of water), completely removed, so that the fossil, which must have originally lain loose in a natural mould of the outer surfaces of the caudal scutes (which has unfortunately not been preserved), exhibits, externally, a complete cast of scutes, and internally, the casts of the outer surfaces and neural canals of a corresponding number of caudal vertebrae.[24]

Here again, the way in which Huxley shapes his narrative is of interest. Though the detail is precise, the organisation is literary. Although he knows at the outset that the paper is about the *Stagonolepis* being in some respects crocodilian, in some respects not, he lets part of his evidence be introduced anecdotally, affecting not to be drawn too rapidly to a conclusion. The word 'speculation' as used above is not intended to be critical in a pejorative sense. Attention is merely drawn to the difference between the extreme precision of the description of the specimen and the relative imprecision of terms like 'appears to have sunk', 'by some agency or other' and 'probably'. Of course he understood this perfectly well, returning in later papers to the problem of why such reptiles as were found in Elgin were not evenly distributed around the globe, and looking for the likely causes of uneven distribution, in order to explain why what in 1859 looked like similar terrestrial conditions did not have preserved in them similar reptiles. But reptile finds were soon to be reported from other parts of the world.

Evidently this article, like many others by Huxley, is a research report, not an attempt at a comprehensive treatment of a subject. He did not deviate for too long, however, from his main subject.

The second peculiarity to which I have directed attention may perhaps be the result of the early anchylosis of the caudal transverse processes with the neural arches. However, this may be, the character in question is a very exceptional one, and long led me to hesitate in regarding the vertebrae in question as really caudal.

The gradual divergence from the strictly Crocodilian type of organization which is manifest in the remains to which I have just adverted reaches its climax in the next part I have to mention. A fragment of bone protruding from one of the blocks of sandstone from Lossiemouth was the last to attract my attention of all the fossils which have been sent me from Elgin. Certain indications convinced me that, notwithstanding the extreme fragility of the bony substance and the depth with which it seemed to penetrate into the sandstone, it was worth some trouble to work out this bone completely, and having succeeded, by dint of careful chiselling, in removing a considerable quantity of superincumbent matrix without damage to the fossil, I was rewarded by the view of the nearly entire ventral face of the coracoid bone, of a form very unlike what might have been anticipated. For this bone, far from having the transversely elongated form more or less constricted in the middle, which is exhibited in the corresponding part in all the true *Crocodilia* with which I am acquainted, whether mesozoic, cainozoic, or recent, is almost elliptical in outline, and the long axis of the ellipse, which is nearly parallel with the middle line, is to the short transverse axis as 6 to about 4. The articular surface for (*sic*) the humerus is crushed and partly broken away, and part of the anterior external edge is incomplete. The posterior edge of the bone presents a deep excavation close to the articular end; and, if two lines were drawn, one longitudinally through the deepest part of the notch, and the other transversely through the greatest transverse diameter of the bone, they would cut one another in the midst of a circular foramen, which corresponds with the coracoid foramen of Crocodiles and many Lizards. I find no coracoid so similar to this as that of *Hylaeosauras*.[25]

Obviously Huxley is trying to relate this creature that he does not know to something that he does know. The fossil will make sense to him if he can so place it. Additionally, he concluded, from his analysis of *Stagonolepis* fragments, that the reptile had shorter and thicker feet than a crocodile, and long claws. These feet and long claws would sometimes have left tracks, but as to the casts of footprints found at Cummingstone, he will only go so far as to say the 'the Crocodilian number of toes, again, combined with the non-Crocodilian proportions of the feet, harmonises very well with the modified Crocodilism (if I may coin a word) of the organisation of *Stagonolepis*'.[26]

It was with this type of paper that Huxley made his reputation amongst his colleagues and peers. Though Huxley had never been to Elgin, Findrassie, Cummingstone or Lossiemouth, there is no room to doubt that the evidence has been firmly examined. The discussion, though cautious, is also firm. Given the fact that the objects being analysed were, at the time, unique, the treatment is altogether fine. Nor had he finished with the subject, since to the printed version of his talk he added a note, which very much shows him 'at work' – that is, very much involved in a continuing enquiry which frequently required new evidence to be taken into account. No immediate criticism of this procedure is intended here. After all, it has always been quite common for scholars to write sets of articles, perhaps leading to a monograph or a book. It is also common for them to take into account comments and reviews as they go along. Often a correspondence arises in some journal or newspaper in which the original writer takes into account, if he can, ideas or objections that have arisen in the minds of others. Admittedly, some careful readers of Huxley's writings have felt that he was too inclined to let what he had written go to press before the research had really been finished. But in the present case, at the beginning of his work on the Elgin fossils, it seems reasonable that he should take any opportunity that offered to add to what he had already said. The addition to the article reads as follows:

> NOTE – Unless the contrary is expressly stated, the preceding paper remains in all essential respects the same as when it was sent to the Society. Since that time, however, several months have elapsed, and, thanks to the exertions of my indefatigable friend Mr. Gordon, much new material has come to light. On the other hand, I have submitted the recent *Crocodilia* to such a revision as the time at my disposal would allow, and I have published some of my results in an Essay 'On the Dermal Armour of *Jacare* and *Caiman*, with notes on the specific and generic characters of recent *Crocodilia*', published in the *Proceedings of the Linnean Society* for February, 1859.
>
> The sum of my conclusions from the various kinds of evidence thus obtained is, that the divergence of *Stagonolepis* from the Crocodilian type is even less than I had imagined; and in some characters, such as the form of the posterior maxillary teeth, *Stagonolepis* is more like a modern Crocodile than a Teleosaurian.
>
> A very fine specimen of a coracoid, recently sent by Mr. Gordon, convinces me that the differences from the Crocodilian type of structure which I have ascribed to this one in *Stagonolepis* do not really exist. I am obliged to take this opportunity of distinctly asserting that the only two caracoids of *Stagonolepis* which have been discovered have been sent to me direct from Elgin, that I have worked them

out from the matrix with my own hands, and that no anatomist had seen the one described in my paper before the publication of my account of its structure. That account, however, is incomplete, the new specimen showing that a considerable part of the bone was wanting, and that when perfect, it is far more Crocodilian than Lacertilian in its characters.

The remarkable new Reptile, *Hyperodapedon Gordoni*, which I have briefly characterized in a note to Sir R. I. Murchison's paper (see p. 435), is of paramount interest; for while, on the one hand its discovery justifies my hesitation in at once ascribing the Cummingstone footmarks to *Stagonolepis*, – on the other, its marked affinity with certain Triassic reptiles, when taken together with the resemblance of *Stagonolepis* to mesozoic *Crocodilia*, leads one to require the strongest stratigraphical proof before admitting the palaeozoic age of the beds in which it occurs.

Finally, I may add, that *Stagonolepis* attained a much greater size than my former materials warranted me in believing. Some of the recent discoveries lead me to suppose that it reached 16 or 18 feet in length.

These paragraphs are quoted in full because they are full of interest. In the first place, he is quite clear that the Elgin rocks in which reptilian fossils were found had to be Triassic. This is belied by the correspondence, however, where he pulls his punches, ascribing his indecision to others' disagreements. Secondly, what he says is enough to indicate that Gordon had sent him more material than is mentioned in the correspondence. Thirdly, we see here the fixation with crocodiles that is represented by various notebooks in the Imperial College archive. In one of these notebooks is the interesting page which is reproduced here – interesting because it shows Huxley working something out negatively by the elimination of possibilities. When other possibilities had been eliminated he was left with crocodiles. But why did he protest that he had done his own work? Presumably he had been challenged on this point, by someone who had questioned the originality of his research. He had not, though, quite done with the subject:

While this Note was passing through the press, the Monograph of D'Alton and Burmeister ('Der fossile Gavial von Boll', 1854) came into my hands. The excellent memoir by these authors on the ancient *Mystriosaurus bollensis* is preceded by a valuable essay on the organization of recent *Crocodilia*, including the best account I have met with of the ventral dermal armour (p. 29). The transverse sutures dividing the ventral scutes, and their mode of articulation, are noted; but, on the other hand, it is said that only the three or four outer series of ventral scutes have pitted surfaces, and the authors suppose that the ventral armour which they describe (and which is apparently that of a *Jacare*) is found in all recent Crocodiles. I can find no reference to the fact that the great majority of living Crocodilia are wholly devoid not only of ventral bony armour, but of *articulated* dorsal scutes. – July 5, 1859. T. H. H.[27]

It would have taken a courageous person to challenge these various conclusions in 1859. Huxley was hardly making large claims, but he was talking about specimens few other people had seen. Time would therefore have to pass before the affinity of *Stagonolepis* to the *Crocodilia* could be properly established or tested.

Gordon had helped Huxley a great deal during 1858 and 1859, as the correspondence shows. Perhaps not quite enough letters have survived for an absolutely precise correlation to be made between what Gordon sent and what Huxley described, yet at this stage Huxley did not have any other source of material. The collaboration was at one of its peaks.

Between December 1859 and August 1866, however, Huxley appears to have heard from Gordon only twice, the problems of the Elgin sandstones having been temporarily shelved while Huxley got on with other things. Letters may have passed between them that are now lost, but nothing of great significance seems to have occurred. During this five and a half year period, Huxley was as usual extremely busy, delivering an important set of lectures on classification at the Royal College of Surgeons, writing and publishing the disruptive *Evidence as to Man's Place in Nature* (1863), delivering the Hunterian Lectures on the structure and classification of mammalia (1864), interesting himself in questions of ethnology and, of course, attending to his varied, often extremely onerous administrative duties.

Of importance to the story being told here is his 1862 paper called 'Geological Contemporaneity and Persistent Types of Life' (see Appendix II) since it shows him continuing to be cautious about ideas of evolution based on fossil evidence. He might indeed champion Darwin when Darwin's ideas were attacked by clerics, but he was, at this stage, still very far from being convinced that palaeontology provided any reason for believing in the successive modification of anatomical structures over a long period of time. It was more in his nature not to believe in modification despite the fact that he used the word. At any rate, he had not himself observed the evidence for it. A letter to Charles Lyell, dated June 1859, probably expresses his restrained, for the most part negative attitude to all the talk about evolution that followed the publication of *The Origin of Species*.

> I do not exactly see the force of your argument that we are bound to find fossil forms intermediate between men and monkeys in the rocks. Crocodiles are the highest reptiles as men are the highest mammals, but we find nothing intermediate between *crocodilia* and *lacertilia* in the whole range of Mesozoic rocks. How do we know that man is not a persistent type?[28]

For better or worse this is typical of the way he often thought. Put up a notion, in this case the notion of a persistent type. No matter that there is no evidence that particularly supports it. Then claim that there does not exist evidence that would refute it. Keep an open mind about the possibility of such evidence turning up. Someone else would have to find that evidence. Huxley's position, when he adopts this strategy, is perhaps logically secure, since he cannot be faulted for remaining unconvinced about evolution until the evidence for it is overwhelming. But, entrenched in a position of his own making, might he in effect be asking for the wrong kind of evidence? In his letter to Lyell, there would seem to be an inherent textual contradiction between the word 'highest', by which Huxley means superior in anatomical organisation or specialisation, and 'persistent', a concept which leaves room only for a very simple idea of superiority.

In the same letter, Huxley says that the transmutation hypothesis would 'harmonize the spirit of Palaeontology with that of Physical Geology', revealing in this remark a long-standing anxiety about the possibility of conflict between palaeontologists and geologists, but not expressing much of a hope about the possibility of reconciliation. He seems more concerned about the reconciliation of the disciplines, or their practitioners (was he being *too much* affected by London rivalries?) than excited by new ideas that might transform both. By comparison with his more open and spontaneous friendship with John Tyndall, Huxley tended to be obsequious in his dealings with Lyell and Murchison. Interestingly, while his work on the Elgin fossils was at a standstill, his thinking about evolution seems also to have been at a standstill. So it remained until 1866 at least and perhaps, in truth, until 1876. In any case, though he would write placatingly to Lyell about the harmonising of palaeontology with geology, it was primarily palaeontological evidence that impressed him.

Work was only at a standstill for Huxley. Other people continued to think about the fossil finds in Moray. If there were reptilian fossils in the Old Red Sandstone around Elgin, might there not also be reptilian fossils in other Old Red Sandstone regions to the north, that is, in Caithness and Cromarty? Alternatively, if the rock in which the fossils had been found was not the Old Red, was it possible that the rocks to the north of the Moray Firth had also been wrongly described? Clearly there was a lot to be puzzled about. The story did not necessarily end with the description of what happened to interest Huxley.

An example of this continuity of research during the period in which Huxley's palaeontology lay dormant (at least as far as his work with Gordon was concerned) is provided by a letter from W. S. Symonds, dated 16 December 1862, which appeared in the *Inverness Courier*. Symonds felt himself to be in a strong position on the very questions that concerned Huxley and Gordon. Unlike Huxley he had had a good look at the region he here wrote about.

> I am one of those 'field geologists' who examined the physical geology and the district, in company with the Rev. G. Gordon and several other geologists, after the meeting of the British Association in Aberdeen, and as I hold *very strong doubts* respecting the assumed Devonian, or Old Red, age of the reptiliferous beds of Spynie, Lossiemouth, &c., perhaps you will do me the favour of inserting the following communication.
>
> All thanks are due to the Rev. G. Campbell of Tarbat, and the Rev. Mr. Joass of Edderton, for their perseverance and good work in having detected the *reptiliferous deposits* at Portmahomack and other localities on the north-west coast of the Moray Firth, although I do not think that any physical geologist, who knows the Elgin district, ever doubted those rocks would in time be detected there. When I visited my friend, Mrs. Hugh Miller, at Cromarty, after our investigation of the Elgin district, I particularly requested her to try and induce her geological friends and acquaintances to search for the reptiliferous deposits in the district around Cromarty, Tain, and other localities. It would give me, Sir, real gratification if the *Telerpeton, Stagonolepis*, and *Hyperodapedon Gordoni* could be proved satisfactorily to be reptiles which were contemporaneous with the deposition of the far-famed old red sandstone of Hugh Miller, or even of the lower carboniferous deposits; but I am compelled to declare, although I cannot enter into particulars in a communication of this kind, that I have no faith in the truth of such a correlation, and I do not think that the discovery of reptiliferous deposits on the north-west side of the Moray Firth has advanced this correlation by a single step. It is my belief that when the very difficult geology of the district in question is worked out by the geological surveyors, or exposed by railways, it will prove to be somewhat similar to the maintops in the neighbourhood of Weston-super-Mare, and other localities of the S. W. of England, where rocks of the age of carboniferous limestone are seen in conjunction with those of the age of the upper trias and lias, and where fossils of palaeozoic and mesozoic character may be obtained almost in contact with each other, as in the railway cutting

> at Uphill, near Weston. In the Elgin district also, it is my opinion, that the reptiliferous deposits, whether Permian or Triassic, are of much later date than the Devonian or old red deposits that they overlie. As regards the stratigraphical reading of the rocks from Portmahomack, by Geanies, to Shandwick and Cromarty, permit me to state, that the only reading which will throw light upon the question, will be the interpretation of those hitherto hidden pages, the *actual points of contact* between the reptiliferous deposits and the underlying old red rocks. The similarity in mineralogical character and colour, between the Holoptychian sandstones and the reptiliferous deposits, will not allow of any other reading in such a difficult district. My brother geologists in the North of Scotland will then, I trust, forgive me if I say that I do not consider that they have as yet proved the Devonian age of the reptiliferous deposits. They have, I believe, great difficulties to contend with, and I am quite prepared to hear that Holoptychian scales have been found in the *same quarry* with reptilian footprints or reptilian scales or bones, but they must bring forward a rock specimen containing, in its matrix, reptilian relics embedded with the remains of Old Red fish; and they must show us a section exhibiting a conformable, upward, continuation from the Holoptychian sandstones into the reptiliferous rocks, before we can accept their present stratigraphical reading of the Elginshire and Ross-shire deposits. When they have done this, no geologist will rejoice more with them than your obedient servant.[29]

Almost the whole of this letter is reproduced here because it bears so directly on an issue that Huxley, despite his private opinions, still in public regarded as unresolved. Gordon acknowledged it on 31 December 1862 [Edinburgh University Library MS 5646] saying that he believed Joass would be able to provide the additional evidence Symonds asked for (not so, of course) and a few days later, Symond's sent both Gordon's letter and his own to the *Inverness Courier* on to Lyell, expressing the view that his letter would have the effect 'of stimulating the efforts of the geologists in the North of Scotland'. These communications represent the sort of thing that was going on while Huxley was pondering the Elgin fossils in London. All these enquiries were soon to be given professional direction by newer members of the Geological Society, and by the Geological Survey.

The Telerpeton

In 1866 came a second breakthrough in the form of the new evidence that allowed Huxley to write the paper called 'On a New Species of *Telerpeton Elginense*', which was read at a meeting of the Geological Society in December and published the following year. Once again one is faced by the paradox of his work in this field. Despite the long span of time devoted to the Elgin fossils, he again rushes to make use of the material Gordon has sent him, so quickly in fact that one can see, in retrospect, that the likelihood of his later having to reconsider his position in the light of other evidence was great. But he had to be first.

As before, he acknowledges his indebtedness to George Gordon in formal terms.

> I am indebted to my friend the Rev. Dr Gordon, of Birnie, by Elgin, for the opportunity of examining the very beautiful specimen of *Telerpeton Elginense* of which I propose to give a description in the present paper. It is the property of Mr. James Grant, General Assembly Teacher, in Lossiemouth, Elgin, who has been good enough to entrust it to Dr Gordon for transmission to me; and it was obtained from the well-known reptiliferous fossil beds of Lossiemouth, along with some highly interesting fragments of *Stagonolepis* and *Hyperodapedon*.[30]

The greater part of this paper is a report on Huxley's minute and painstaking examination of the parts of the specimen sent from Elgin, the double conclusion showing a greater firmness of direction that he had found possible only a few years earlier. He says, first:

> From the description of the organization of the *Telerpeton Elginense* which has now been given, it is obvious that this animal is one of the *Reptilia*, devoid of the slightest indication of affinity with the *Amphibia*. It is Saurian in all its characters; and if we enquire to what division of the Sauria *Telerpeton* belongs, there appears to me to be no doubt that it must be referred to the true *Lacertilia*, and among them to the sub-order *Kionocrania* of Stannius, which contains all the modern Lizards – though I cannot make sure, from the present specimen, that it possessed a columella.[31]

What Huxley knows about is the anatomical structure or organisation, as he continues to call it in this case of reptiles. He has acquired this knowledge by a patient study of all the actual specimens he could contrive to see and, also, as the footnotes to this page show, from other people's articles, even when it was impossible for him to see what he was describing. Still trusting the verbal descriptive method he and most of his contemporaries had had to adopt, Huxley makes solid use of this compounded, accumulated knowledge, in the process of doing so demonstrating the power of his specialised approach.

The second part of the conclusion to the article on the Elgin *Telerpeton* reverts to the notion of persistent types.

> It is most interesting to observe that *Telerpeton* presents not a single character approximating it towards the type of the Permian *Protosauria*, nor to the Triassic *Rhynchosaurus* and other (probably Triassic) African and Asiatic allies of that genus, nor to the Mesozoic *Dinosauria*; still less can it be considered a 'generalised' form, or as, in any sense, a less perfectly organised creature than the Gecko, whose swift and noiseless run over walls and ceilings surprises the modern traveller in warmer climates than our own. And whether the age of the deposit in which it occurs be Triassic or Devonian, *Telerpeton* is one of the most astonishing examples within my knowledge of a *persistent type* of animal organization.[32]

This concluding paragraph of the article called 'On a New Species of *Telerpeton Elginense*' shows the true Huxley of 1867. He now knows much more. He can place his study of particular fossils within a much wider context – Africa, Asia. He is confident about his own work. As for the age of the rock in which this fossil was found, he continues to defer to the geologist, no doubt knowing that people sitting in his audience had yet to reach agreement about the Elgin Sandstones. He keeps well away from stratigraphical disputation, continuing to show no interest whatsoever in the Elgin terrain. He could do this because of a strengthening conviction that palaeontology was an absolute and self-sufficient form of knowledge. If the anatomical organisation of a vertebrate fossil could be described, its 'place in nature' would be known and no other knowledge would be required for an understanding of it. He moves directly from description to classification without looking either to left or right, or at least without any stated anxieties about the possible limitations of his method. What helps him to do this is the still firmly entrenched concept of persistent types – that is, types of anatomical organisation that have persisted over a long period of time without significant modification. (Again, see Appendix V).

Where this concept is presented in a simple form, as it is here in the *Telerpeton* article, Huxley perhaps gives the impression of begging a number of important questions. The first has to do with time. What, in this context, is a 'long period of time'? If persistent, then persistent for how long? It would eventually be understood that there had been two generations of reptiles, both now extinct, in the Elgin region, some still flourishing 220 million years ago, the others 250 million years ago, Huxley's being in the latter group. Both were in the main Triassic and both were rendered globally extinct by some event or set of events even now not securely identified. But one looks in vain in Huxley's article for some estimate of the length of time the right conditions had prevailed for those reptiles to continue to live. At least in this particular set of papers, the *Telerpeton* piece being one of the set, he refrained from making a correlation between the physical makeup of the creatures he described and the conditions necessary for their survival. Consequently the extent of geological time was not an important issue for him. On the contrary, his taxonomic preoccupations remained dominant. A second question concerns modification. In what circumstances and by what mechanisms might modifications to anatomical structure occur? Here, too, it can be seen that as far as his own work was concerned Huxley was satisfied by small advances, was prepared to postpone committing himself to conclusions about widely discussed issues until he was presented with factual evidence of a kind acceptable to him, and was willing to write an article that was descriptively strong though intellectually extremely cautious. The article being discussed here is still work in progress; if it is work in progress, tentative conclusions are perfectly respectable. Perhaps it might be fair to observe that the rigidity of Huxley's method, despite or perhaps because of, the rigour of its application, functions here as an imprisoning cantonment preventing or disinclining him from thinking about what he must have appreciated were important aspects of his own subject.

The Hyperodapedon

Huxley's progress towards the completion of his work on the Elgin reptiles was marked by a third paper entitled 'On Hyperodapedon' (1869). Once again, there will be an advantage in allowing Huxley to speak for himself.

> A little more than ten years ago, namely, on the 15th December, 1858, Sir R. Murchison read a paper 'On the Sandstones of Elgin' before this Society. It was followed by an essay of my own, 'On the *Stagonolepis Robertsoni*' , an animal so named by Prof. Agassiz in his *Poissons fossiles du Vieux Grès Rouge* from some impressions of its dermal covering which had been discovered in the Elgin Sandstones. In the latter paper, and in notes added to both papers, before their publication in the middle of the following year, the fact that *Stagonolepis* was a reptile closely allied to the Mesozoic *Crocodilia*, though distinct from any known form of that age, was for the first time asserted, and, indeed, I may say demonstrated, the remains placed in my hands enabling me to put the fact beyond doubt. At the same time I mentioned the existence in the same beds of a 'Saurian reptile about 6 feet long, remarkable for the flattened or slightly concave articular surfaces of the centra of its vertebrae, and for its well-developed costal system and fore and hind limbs, but more particularly characterized by its numerous series of subcylindrical palatal teeth'. I named this new reptile *Hyperodapedon Gordoni*, in

honour of its discoverer, the Rev. Dr Gordon, to whose exertions in the Elgin country geology owes so much: and I stated that 'its marked affinity with certain Triassic reptiles, when taken together with the resemblance of *Stagonolepis* to Mesozoic *Crocodilia*, led one to require the strongest stratigraphical proof before admitting the Palaeozoic age of the beds in which it occurs'.

Sir R. I. Murchison admitted that his belief in the Devonian age of the Elgin sandstones was 'somewhat shaken' by the discovery of the nature and affinities of these reptilian remains.

In the ten years that have elapsed since the papers to which I have referred were read before the Society, the age of the reptiliferous sandstones of Elgin has been repeatedly discussed by some of the most eminent of English geologists, with the general result that while one half of the disputants produced excellent reasons for believing them to be of Mesozoic date, the other half adduced no less weighty arguments in favour of their Palaeozoic age. And it is a curious circumstance that in this Geological Siege of Troy, Priam had been fighting the battle of the Greeks, and Nestor that of the Trojans, – Sir R. Murchison, whose general geological views would naturally incline him to assign a later date to these Elgin Reptiles, having been the sturdiest champion of their Devonian age, while Sir Charles Lyell who ought to rejoice if they could be made out Palaeozoic, has as strongly fought for their belonging to the Trias. Without meaning to compare myself to Achilles, I may say that 'under these circumstances' I thought it best to retire to my tents and take no part in the fray until my palaeontological armory should yield more efficient weapons. As my excellent friend Dr Gordon supplied me from time to time with new specimens, I lived in hope that one day or other I should be able to make an effective sally.

No such opportunity presented itself, however, until the year 1867, when a number of important facts came to light in singular coincidence, and, as I conceive, rendered the proper discussion of the question and the drawing of satisfactory conclusions easier than before'.[33]

The facts to which Huxley refers here are new specimens, two of them coming from the Triassic sandstone of Coton-End in Warwickshire, several others from Lossiemouth and others (that had been in the possession of the Geological Society without Huxley's knowing it) from India. This permitted the type of comparison that Huxley was so good at, established at last the age of the deposits from which the Elgin reptiles had been taken (though those deposits still had to be described accurately) and resulted in Murchison's conceding that he must have been mistaken about the age of 'overlying sandstones and limestones in Moray'. The fossil evidence had resolved a problem that had puzzled them all for a decade.

This 1869 *Hyperodapedon* paper is worth close attention because it so clearly reveals Huxley at work. As the correspondence also shows, he had for ten years been extraordinarily patient or, perhaps, hesitant even, despite his tendency at other times to rush into print. It had been on 23 April 1859 that he thanked Gordon for the 'last shipment'. By May of that year he had been able to say it had some relation, he thought, with *Rhynchosauri*, and he said so again in the letter of 11 November 1859. But then eight years passed until, in 1867, he began to be aware of the evidence referred to in his preamble quoted above – eight years that are pretty important in the history of British palaeontology. This evidence allowed him to describe a previously unknown creature of considerable strength and bulk, characterised by clawed feet, powerful legs, a heavy head, distinctive jaws and a curved beak. A first. The measured, professional tone of the paper did not disguise from his audience that something special was being described. On this occasion at least patience had been rewarded.

The article itself demonstrates both the extent of Huxley's reputation as a palaeontologist, since people turned naturally to him when vertebrate fossils were being identified, and also the growing awareness of the subject which resulted in scientists in different places pooling their knowledge under the aegis of the Geological Society. Huxley acknowledges not only the help of George Gordon, but also that of Mr. Grant, the teacher at Lossiemouth who had sent him the new specimen, a Mr. Lloyd who shared with him specimens that had been collected earlier in Warwickshire, Professor Oldham, Director of the Geological Survey of India, and a Mr. Whitaker, whose researches had been conducted in Devonshire. It is fashionable in some quarters to criticise Victorian scientists adversely as hungry for class power, in as far as their institutions consolidated their social or even, some would say, imperial position, allowing them to exploit the labour of others while achieving for themselves a privileged metropolitan status. In the case of Huxley it is true that information flowed *to* him. It is also true that he benefited from Imperial technology; he could *expect* to know what was happening in other parts of the world. But this granted, surely it also has to be acknowledged that the mechanism of the Geological Society, with its meetings and publications, worked – and worked well. It is difficult to imagine that palaeontological enquiry could have advanced as rapidly as it did without the help of a central institution of one kind or another. The second two pages of the *Hyperodapedon* paper provide an admirable example of the type of collaboration that, if not impossible, would have been much more difficult to achieve without the Geological Society.

Huxley comments on the peculiar interest of a discovery that so vastly extended his knowledge of *Hyperodapedon* and its global distribution. Like the *Stagonolepis* when first described by Huxley, the *Hyperodapedon* sent from Elgin was a unique specimen. But this situation did not last for long. By 1869 Huxley was able to make anatomical or structural comparisons on the basis of several partial fossils he had by that time been sent from various sources. He was in his element constructing ideas out of others' field work.

These comparisons allowed him to identify the special features of the *Hyperodapedon* jaw.

> On looking at the original specimen, it became clear that this groove corresponded with the depression into which the oral edge or the mandible is received when the mouth is shut. The opposed faces of the palato-maxillary teeth had been worn flat by attrition against the opposite sides of the mandibular teeth, which work between them as a knife-blade shuts into its handle; and it followed that the dentary margin of the mandible must be worn to an edge adapted to fit into the groove. So far as I know, no other fossil reptile possesses any such peculiarities; and thus this interesting fragment presented me with new means of distinguishing the teeth and jaws of *Hyperodapedon* from those of other Reptilia.[34]

Huxley did not illustrate this in the 1869 article, allowing the words of the paragraph quoted above to do their own work, but he did illustrate it when he returned to the subject in 1887. On that later occasion he described the *Hyperodapedon* jaw in the following terms:

> In *Hyperodapedon*, the mandibular rami are extremely massive, and are ankylosed together in front, throughout their long symphysis. The anterior and superior angle of each is prolonged into a 'rostral' process, which passes upwards, outwards, and forwards. The two thus leave a wide V-shaped interval, into which the end of the premaxillary beak is received when the mouth is shut. These mandibular rostral processes were probably clothed, like the premaxillary rostrum, with horn; and, with the interlocking premaxillary beak, must have constituted a pair of pincers of great strength. Behind this forked mandipular beak, the upper contour of each ramus is strongly concave upwards, in correspondence with the maxillo-palatal convexity, and its close-set teeth are worn to a sharp scissor edge, which fitted into the fore part of the mandibular groove of the upper jaw.[35]

By 1887 he knew more, so he could write a more firmly worded description. We have here a typical piece of Huxley research. He heard about the *Hyperodapedon* in 1858; he waited until additional evidence came to him; he published an article in 1869 based on important but limited information, an article which showed, however, how good he was at analysing partial specimens, even of previously unknown creatures; and in 1887 he returned to the subject (thirty years later at the age of 62) to endorse, qualify and consolidate his earlier research. It could well be said that many scientists, and not only scientists, of necessity proceed in this way, since research never ends, new knowledge always has to be taken into account, and supposedly definitive or absolutely 'true' statements are often invalidated by later work, to the extent that it is as important to appreciate the instability of scientific knowledge as absolutely to believe in it at any given time. Nonetheless, it is correct to notice that Huxley was often extremely tentative, as can be seen in this paper where, particularly, the 'discussion' which constitutes about a third of the whole, is of considerable interest because of what it tells us about Huxley's habits of thought.

The first question is that of taxonomic placement. Where exactly did *Hyperodapedon* fit in? To what other known creatures was it most closely related?

> With respect to the affinities of *Hyperodapedon*, there can be no doubt that it is very closely allied to the genus *Rhynchosaurus* established by Prof. Owen upon a fossil skeleton from the Trias of Shropshire. But *Rhynchosaurus* has shown no sign of teeth in either upper or under jaw, and seems to differ from *Hyperodapedon* as *Oudenodon* does from *Dicynodon*. There is a certain analogy with another Triassic genus, *Placodus*, in the dentigerous roof of the palate of *Hyperodapedon*; but too little is known of the rest of the organization of *Placodus* to test the value of this analogy.
>
> I do not find grounds for assuming any special affinity between *Hyperodapedon* and *Telerpeton*; nor has *Hyperodapedon* anything to do with the Thecodonts, Crocodilian or otherwise, which abound in the Trias.
>
> But it is a very remarkable circumstance that it is nearly allied to an anomalous lizard, *Sphenodon (Hatteria)* which still lives in New Zealand.[36]

Leaving the New Zealand *Sphenodon* aside for one moment, one might perhaps with later judgement argue that Huxley had not been able to carry the question of placement very far in this passage. But it nonetheless shows the better side of Huxley. He acknowledges his indebtedness both to Owen and to Dr Gunther who had written about *Sphenodon* in the *Philosophical Transactions*. Secondly, he begins to make sense of a much wider range of data than he had handled before. Thirdly, his rhetoric leaves his audience in no doubt that the subject had been dealt with judiciously. This was truly pioneering work, and whether right or wrong it would be seen to be pioneering work of importance.

In the discussion part of the paper, Huxley reverts to the question of persistence.

> Moreover it has been seen that *Hyperodapedon* is nearly allied to the living lizard *Sphenedon*, as nearly, I am inclined to think, as to its Triassic congener *Rhynchosaurus*. And if this extraordinary form has persisted with so little modification from the Trias till now, why may it not have inhabited the dry land of the Permian, of the Carboniferous, or of the Devonian Epoch?[37]

He says that, even though the much discussed problem of the age of the Elgin sandstones has now been settled (in his mind at least), to the extent that he knows the *Stagonolepis*, *Telerpeton* and *Hyperodapedon* were all, in that district, found in Triassic rock, he cannot close his mind to the possibility that the same reptiles might have existed in other eras. As usual he confines himself strictly to what he knows. But he also, and reasonably enough, continues to be impressed by persistence over long periods of time – 'reasonably enough' because he has not formulated a notion of what a long period of time might be, and is very far from perceiving how modification might be effected, and in how many generations. He resorts to the coinage of 'Poikilitic' (borrowed from Conybeare) to refer to the extended period of time in which *Hyperodapedon* and its relatives might have lived. Provocatively, he says:

> in speaking of the Poikilitic period, I should like to make its earlier and later boundaries as hazy as possible, and to apply it exclusively to terrestrial conditions and to land and freshwater fauna, without prejudice to the limits in time of the marine conditions known as Permian and Triassic.[38]

The position Huxley adopts here is logically difficult to fault. He gives a clear account of what he regards as factually certain. In matters he believes are uncertain, he makes it quite clear that he has an open mind. No wonder that he influenced others, who aspired to the same clarity. Nor is he intellectually at a standstill. If the distinction between the terrestrial and the marine may seem wrong-headed (at least as far as stratigraphical terminology is concerned), it can be seen nonetheless that Huxley is beginning to direct his attention, not just to persistence, but also to the conditions that permitted survival. If one could not say how a species originated, perhaps one could think about how it died out. He had moved from the local to the global. This was an advance of thought (he still lacked ecological evidence) which left him in a more secure position than, say, *The Cambridge Field Guide to Prehistoric Life*, where one finds the cavalier statement that *Rhynchocephalians* 'evolved and spread in the Triassic period, then largely fizzled out'.[39] Fizzled out! Given the seriousness and difficulty of all discussion about the extinction of species what kind of term is that? Interestingly, however, although Huxley is on the move, one sees once again in this passage that he vastly prefers the security of logic, even when deployed negatively, to the dangers of speculation and the restlessness of enquiry. An unkind critic might say that he had raised the question of distribution but not done very much to provide an answer. And it must be mentioned in passing that the allegiance between *Hyperodapedon* and *Sphenodon* has in recent years been severely challenged.

Huxley did not use the word 'evolution' in the 1869 *Hyperodapedon* paper. His instinct still was to arrange and classify what he knew in terms of structure, anatomy, organisation – and now, albeit in a very simple way, ecology. As far as evolution was concerned he was at this stage prepared to go no further than what he must have considered to be the challenge of his concluding paragraph.

> So long as mammals and birds were known to occur no further back than the older Tertiaries or the middle Mesozoic rocks, it might be legitimate to imagine that they came into existence somewhere between that time and the end of the Palaeozoic series. But now that both can be traced back to the Trias, that it is known that the Crocodilian and Lacertian types of reptiles were then in existence, and that the Amphibia were elaborately represented, I confess it is as possible for me to believe in the direct creation of each separate form as to adopt the supposition that mammals, birds, and reptiles had no existence before the Triassic age. Conceive that Australia was peopled by kangaroos and emus springing up ready-made from her soil, and you will have formed a feat of imagination not greater than that requisite for the supposition that the marsupials and great birds of the Trias had no Palaeozoic ancesters belonging to the same classes as themselves. The course of the world's history before the Trias must have been strangely different from that which it has taken since, if some of us do not live to see the fossil remains of a Silurian mammal.[40]

This rhetorical conclusion, with its bow in the direction of Sir Roderick Murchison, must be interpreted to mean little more than that Huxley's mind was open to ideas of continuity. He could still more easily imagine species persisting than evolving. He has on this occasion nothing to say about how kangaroos came into existence if not ready-made. As palaeontologist he can contemplate the possibility of someone else discovering Silurian vertebrate fossils; he has no inclination to do the work himself. The 'if' of the last sentence protects him from having to carry the subject forward. This was his way. The rhetoric that was, and is, so often admired, allowed him always to make firm statements that were difficult to challenge, though they also date him, since such rhetorical flourishes would be unusual these days in a scientific paper. Were not his peers who participated in the discussion of this paper in 1869 perhaps overly polite?

Interestingly, when, in that discussion, he was challenged on his use of the word 'poikilitic', he replied that

> in using the term 'poikilitic' he was desirous of indicating that while several marine formations with changing forms of life succeeded each other, the terrestrial forms may, in certain cases, have been continuous. He believed that terrestrial forms were at least as persistent as marine.[41]

These are obviously not the remarks of someone who has come to terms with Darwin. On the contrary, they are the remarks of a cautious, somewhat hidebound anatomist in the school of Cuvier.

A few years ago, M. J. Benton conducted a very thorough study of the *Hyperodapedon*, which he published as 'The Triassic reptile *Hyperodapedon* from Elgin: functional morphology and relationships', – a paper which is so comprehensive in its treatment of the subject, and also expertly illustrated, that there is neither room here nor need for any modification of what he said. By comparison with the perhaps half dozen specimens that Huxley eventually inspected, Benton lists over forty, estimating that all the fossil remains taken together might represent as many as thirty seven individuals, most of them from Lossiemouth. These forty specimens had been dispersed. Four are in the Elgin Museum, nineteen in the British Museum (Natural History), seven in the Royal Museum of Scotland, and the others in the Museum of the Geological Survey, the Glasgow Museum at Kelvingrove, the Manchester Museum and the Geology Department of Newcastle University. The list shows that George Gordon was one of the more vigorous collectors. In 1858, he gave his first three, perhaps four, finds, to the Elgin Museum where they still are, and of these the first two formed the basis of the 1869 Huxley paper which is being discussed here, – i.e. in the Elgin system of enumeration, EM 1978.566.1,2 and EM 1978.567. 1–3. In 1885 he donated to the British Museum at least five specimens, which Huxley used, at least in part, for his second *Hyperodapedon* paper. Huxley had worked wonders in 1869 with the materials at his disposal (a skull and a separate partial skeleton), but inevitably made mistakes, some of which Benton gently identifies.

Benton describes the *Hyperodapedon* as a squat, successful rhynchosaur, between four and five feet in length and a specialised skull. It had 'a large nasal capsule and a large eye with sclerotic plates', suggesting a good sense of smell and diurnal vision. In Benton's synopsis, the following paragraph is of particular interest in the context of the Huxley–Gordon correspondence.

> Evidence from the jaw articulation and tooth wear shows that *Hyperodapedon* had a precision-shear bite. Food was probably scraped up by the premaxillary beak and manipulated by a powerful tongue. The batteries of teeth on maxilla and dentary retained open roots throughout their functional life with no normal reptilian replacement. The hind-limb was adapted for scratch-digging, from a comparison of its claws with those of modern mammals. The diet probably consisted of tough, but not hard, plant material.
>
> The forelimb may have been held in a semi-sprawling pose with humerus rotation important in locomotion. The hind-limb was adapted for a semi-erect gait and the femur moved back and forwards with little rotation.
>
> *Hyperodapedon* is found in aeolian sands, and its skeletons show little physical disturbance, though there is evidence of scavanging.
>
> The late Triassic rhynchosaurs from Elgin, Brazil, and Argentina form a close taxonomic group quite distinct from those of the middle Triassic, on the basis of univariate and multivariate analyses. The Indian form is returned to the genus *Hyperodapedon* as *H. Huxleyi* since it is so similar to the Elgin form. Hyperodapedon differs from the South American *Scaphonyx* in the presence of lingual teeth on the dentary.[42]

This 1983 statement by M. J. Benton will perhaps allow the present day reader to see how Huxley, on the basis of limited evidence, began a descriptive and analytic process whose career can be arrested, as it were, in the late-twentieth century by reference to a much greater, indeed global set of specimens.

In order to establish what was the type specimen of *Hyperodapedon gordoni*, Benton naturally scrutinised Huxley's work with the greatest care, attempting to match what Huxley said in the 1869 article with the specimens in the Elgin Museum. His purpose was more to establish the status of former descriptions than to assess the way in which Huxley worked, least of all in a negative hyper-critical spirit. Nevertheless, the four paragraphs in Benton's monograph entitled 'The Holotype of *Hyperodapedon gordoni*' are so relevant to the present study that they are quoted here verbatim.

The holotype of Hyperodapedon gordoni

Huxley (1859, 1869) regarded the two individuals EM 1978.566 and EM1978.567 as the single holotype specimen. A brief review is necessary in support of the lectotype designation given here.

Huxley's original note (1859, p. 435) mentioned one specimen showing vertebrae, ribs, fore-and hindlimbs, and skull. He noted the concave articular surfaces of the vertebrae and the 'numerous series of subcylindrical palatal teeth'. All of these features are seen in EM 1978.566.

In his first full description of *Hyperodapedon*, Huxley (1869, pp. 142–144) described 'the specimen on which the genus was founded'. he stated that 'the remains of the specimen are exhibited by the opposed faces of broken blocks of sandstone, some of which have been separated by splitting along the plane in which the fossil lay' and went on to describe (pp. 141–142) the 17 vertebrae and ribs, clearly EM 1978.566.2 ('block No. 1'). There is no evidence that a counterpart was collected, although it is implied by Huxley's remarks. Further, on p. 142, he stated that the 'abdominal ribs – are better shown in another slab' (probably EM 1978.567.2; figure 3). Huxley (1869, pp. 142–143) went on to describe EM 1978.566.1, containing 'the anterior cervico-dorsal vertebrae and the skull – bent round so that its axis is nearly at right angles with that of the body'. Much of the description seems to suggest that the skull was better preserved in 1869 than now (e.g. the 'downwardly curved, beak-like anterior termination of the snout, which appears to be formed altogether by the premaxillae' (p. 143) is no longer present). The remaining description of limbs and girdles (pp. 143–144) refers mainly to EM 1978.566.2 ('block No. 1'). Huxley (p. 143) referred to a cast of the 'scapula' of 'block No. 1' on 'block No. 2' which also showed 'sundry scattered imperfect impressions on limb-bones', probably the broad anterior dorsal ribs, partial humerus (?), and hindlimbs of EM 1978.567.2. The final comments on teeth (Huxley 1869, p. 144) may refer partly to EM 1978.566.1 and partly to EM 1978.567.1. In later accounts, Huxley (1887, p. 675) continued to refer to a single 'type specimen, now in the Elgin Museum'.

It is clear that Huxley regarded EM 1978.566 and EM 1978.567 as counterparts of one specimen. This is surprising since the former is preserved on its side, and the latter in a dorsoventrally flattened orientation, there is no way that the two specimens can be matched, each has similar portions of the skull preserved, and EM 1978.567 is about

> three-quarters of the size of EM 1978.566. Every bone is numbered in red paint, and some bones are outlined in red, all most probably done by Huxley himself (A. D. Walker, personal communication, 1981), and the blocks remaining are 1A–C, 1E (11B), 2(EM 1978.566.1,2) and 3, 3′, 3″, 3‴, 4, 4′, 5, 7, 8 EM 1978.567.1–3). It seems most likely that George Gordon sent all of the blocks of EM 1978.566 to Huxley in 1858, and may have sent EM 1978.567 then or later. By 1869, Huxley had numbered all the pieces and considered that they all belonged to one individual. However, the original brief description, and the bulk of the 1869 account clearly refers to EM 1978.566, and this specimen shows most anatomical features. It is designated as lectotype with EM 1978.567 as paralectotype.

In a civilised, professional and helpfully exact manner, Michael Benton is drawing attention here to some of Huxley's errors – errors which result in part from Huxley's method, and in part from the limited number of specimens he was able to examine. Because Benton knows more, he can not only confirm the type specimen, but also more reliably describe what Huxley might have called the *Hyperodapedon*'s place in nature. It will be remembered that Huxley used terms like 'allied to', 'affinity' and 'analogy' when he attempted to establish the relationship of the *Hyperodapedon* to other known reptiles. These terms, harmless in themselves, were consistent with his static, non-evolutionary approach, since he did not yet have in mind a line of descent, or see the need for one. In his later publication, *The Reign of the Reptiles*, Dr Benton proposes an evolutionary 'tree' in which, under the heading 'Diverse Diapsids of the Triassic', he calls rhynchosaurs a 'close relative' of the archosaurian family, and the *Hyperodapedon* a 'typical late Triassic form' of rhynchosaur. At issue is the question of which anatomical features can most reliably be utilized to distinguish between, and therefore classify, the fossil remains of extinct reptiles, and an important part of the answer to this question had to do with the structure of the feet and ankles. Huxley was probably aware of this but, with the evidence available, could carry the matter only so far. The question is a technical, biomechanical one and a person wishing to pursue it could hardly do better than begin with the book mentioned above.

As far as Huxley's correspondence with George Gordon is concerned, it is clear that intellectually they had by this time parted company. Huxley had always (since 1859) understood that, if *Stagonolepis* were a reptile, the Elgin deposits had to be Triassic. During the same decade Gordon had unscientifically hoped that an earlier date would make Elgin unique and famous. Famous it had become, but it was not unique. Ten years had passed in which Huxley had deferred to friends and colleagues who were geologists; it was *their* job to settle questions of stratigraphy (though why he was prepared *so much* to defer to them is a professional and psychological mystery any biographer would need to think about before attributing to Huxley bull-dog like qualities).

At all events the correspondence shows Gordon continuing to write about matters that no longer interested Huxley.

In 1875, Huxley returned to the description of *Stagonolepis Robertsoni*, seventeen years having passed since he first began to think about this set of fossils. The new article had the revealing title 'On Stagonolepis Robertsoni, and on the Evolution of the Crocodilia'. By this time he believes he is in a position to describe authoritatively the relationship of the *Stagonolepis* to the *Crocodilia*, which he divides into three sub-orders, the Parasuchia, the Mesosuchia and the Eusuchia, *Stagonolepis* being a genus of the first sub-order.

First for *Stagonolepis* itself. Thanks to George Gordon whose help is again acknowledged, he has now seen more specimens, which he lists at the outset as:

> (i) the dermal scutes which formed a dorsal and a ventral armour;
> (ii) vertebrae of the cervical, thoracic, lumbar, sacral, and caudal regions;
> (iii) ribs;
> (iv) parts of the skull and teeth;
> (v) the scapula, the coracoid and the interclavicle;
> (vi) the humerus and (probably) the radius;
> (vii) the ilium, ischium, and pubis;
> (viii) the femur and (probably) the tibia;
> (ix) two metacarpal or metatarsal bones.[43]

Once these specimens have been listed as a means of indicating his new position, he devotes the next part of the paper to precise descriptions of them, with a view to relating *Stagonolepis* to other creatures of the same geological age, or in the same line of descent. In this exercise he comes to the conclusion that

> there can be no doubt that *Stagonolepis* and *Beledon* are members of one and the same natural group, and that this group must be included among the order *Crocodilia*.

Having established these relationships to his own satisfaction, he then introduces a working definition of the *Crocodilia*, a definition which it seems not necessary to quote here.

As already indicated, various notebooks in the Imperial College archive attest to Huxley's long-term interest in crocodiles, an interest that extended over the whole period covered by the Huxley–Gordon correspondence. Of particular importance is Volume 92 of the Huxley archive, which amongst other things contains early clippings from issues of the *Quarterly Journal of the Geological Society* (including one from a piece by Mantell), notes on *Stagonolepis Robertsoni* dated 26 November 1858 (the beginning of his serious research?) and an entry indicating that a set of specimens had been returned to Elgin by the end of that month (showing, perhaps that the drawings had already been made for his first *Stagonolepis* article).

On folio 14 of the 'Fossil Crocodilia' volume we find Huxley listing what, on the evidence of the vertebrae, *Stagonolepis* is not, after which he says that it may be

(i) Crocodilian
(ii) Dinosaurian
(iii) Thecodontosaurian

In these notes to himself, he is working by a process of elimination. Reasonably enough, considering that his specimen was unique at the time the notes were made. Later on in this notebook he refers both to a German work on crocodiles by Brown and Kamp, and to Cuvier's notes on the 'Crocodile de Caen' in the tenth volume of Cuvier's collected works.

This notebook obviously represents, at least in large measure, Huxley's first work on the *Stagonolepis*, but his enquiries over the years were continuous; the *Lacertilia* drawings, for example (Imperial College E 48), are dated 1875. It can reasonably be supposed that he once again took up the question of the relationship of *Stagonolepis* to the *Crocodilia* as he recovered from his illness, that is, in 1873 or so. All the time he had had his sights set on the Monograph he had promised so many years earlier; both the notes, some of them in Imperial College, and the 1875 *Stagonolepis* article, represent steps towards that goal. Indeed, much of the 1875 article is incorporated more or less verbatim into the Monograph, including the definitions of the sub-orders of *Crocodilia* referred to above. Another person would have waited until 1877 to make a major, perhaps final statement about *Stagonolepis*, but that was not Huxley's way. For reasons peculiar to himself, he felt obliged to report his interim findings to the Geological Society in 1875, a strategy which also allowed him to take people's responses to his ideas into account.

What about the idea of evolution in Huxley's 1875 title 'On *Stagonolepis Robertsoni*, and the Evolution of the Crocodilia'? To what extent had he altered his position on this important subject since his illness? Had he at all modified his ideas about persistent types?

Of course it is important to pay close attention to Huxley's own words; anyone really wishing to come to terms with the question will wish to have Huxley's article at hand. Having, distinguished between *Parasuchia* (of which he now wrongly believed *Stagonolepis* was a member), *Mesosuchia* and *Eusosuchia*, these distinctions having principally to do with crocodiles and only peripherally with the Elgin reptiles, he refers to what he calls the 'complete ancestral tree of the *Crocodilia*'. By this he means that in sets of creatures whose anatomical structures are more striking to him in their similarities than in their differences, the differences nonetheless allow him to establish degrees of descent, and this in terms of measurable characteristics irrespective of any other factors. And one has at this point to be careful not to misrepresent him since, by the standards of his time, he is being extraordinarily thorough in noticing, measuring and taking account of those differences, whether as zoologist or palaeontologist. He knows as much about crocodiles as any Briton; his knowledge has been tested in the lecture room, as well as in learned journals; and he has read widely in the foreign 'literature' of the subject. Full credit must be given to the energy and determination with which Huxley conducted the study of the crocodile, whatever the accuracy of the ancestral tree. At the turn of the discussion he says:

> The results obtained from the investigation of the nature and order of the successive changes which the *Crocodilia* have undergone since their first appearance in the Triassic epoch, naturally suggest the enquiry whether the nearest living and extinct allies of the Crocodiles, the Lacertilia and the Ornithoscelida, which are traceable, the former from the present day to the Permian epoch, the latter from the later Cretaceous to the Triassic epoch, exhibit any evidence of having been subjected to a similar process of evolution.[44]

Three paragraphs later he adds

> that it would appear that, on the whole, the vertebral system of the Lizards has undergone a change corresponding to that which has occurred in the Crocodiles, and that this modification of the articular faces of the vertebrae took place at an earlier period in the Lizards than in the Crocodiles. Apart from this there is no evidence that the Lacertilian type of structure has undergone any important change from the later Palaeozoic times to the present day.[45]

When Mario di Gregorio discusses Huxley's work on crocodiles in his *T. H. Huxley's Place in Natural Science*, he remarks that 'Huxley had made a decisive step towards an evolutionary interpretation of the crocodilian forms', because 'if palaeontological evidence can be found to show that structurally more primitive forms are *temporally* more ancient than the more complex, this will be good evidence of the occurrence of evolution',[46] and generally he writes as though Huxley was by this time more fully committed to the concept, saying that 'Huxley's survey of the Crocodilia is an early and interesting case of a survey of the evolution of groups above the level of species'.[47] Some difference of opinion is only to be expected in this kind of discussion. Huxley as usual has been extremely positive in his descriptions and comparisons, but cautious in his general remarks. Mario di Gregorio's paraphrase of Huxley's argument is equally positive, but also, as far as general assessment is concerned, cautious, since to talk about the 'occurrence' of the idea of evolution in 1875 is not to talk about very much. The view of the present writer is that Huxley was not yet fully persuaded. His work was not over. He knew it was not.

Besides this, the 1875 paper is not wholly coherent either in logic or presentation, a fact which frequent commentators on Huxley's work, like Professor Seeley, were quick to point out in discussion. Seeley 'saw no ground', it was reported, 'for doubting the value of Professor Huxley's conclusions with regard to *Stagonolepis*, but thought that it might be to the advantage of science to have other ways of looking at the evolution of the Crocodilia'.[48] In his reply to Seeley, Huxley simply said that 'the existing Lacertilia were little modified, and the existing Crocodilia much modified, descendants from a common ancestral form of palaeozoic age'.[49] With Owen's earlier work in mind, Seeley was surely not surprised to see Huxley evade the implied challenge. Perhaps then it would be accurate to say that in 1875 (and perhaps indeed since 1870) Huxley had come to see that fossil evidence *might* be consistent with what he still called the 'hypothesis of evolution'. His contribution to the subject was to take note of skeletal or anatomical differences and similarities, but as palaeontologist he was unable to participate in any discussion of the processes of modification or adaptation.

The Monograph

Huxley was now ready to write what became *Monograph III* of the Memoirs of the Geological Survey of Great Britain, the very thing he had promised to do in 1859, shortly after he had made his first contact with George Gordon and had his first insight into the fact that *Stagonolepis Robertsoni* was of more than local significance. His new title incorporates the conclusions to which he has come. It has become: 'The Crocodilian Remains found in the Elgin Sandstones, with Remarks on the Ichnites of Cummingstone'.[50]

In the Introduction, Andrew Ramsey, by then Director General of the Geological Survey of England and Wales, drew attention to the importance of Huxley's research.

> The masterly work done by the Author in proving the crocodilian nature of *Stagonolepis*, taken in connexion with the discovery of *Telerpeton* and *Hyperodapedon* in the same strata, very soon satisfied Geologists that the Elgin Sandstones must be considered part of the Trias, and, most probably, to that part of the series known as the Lower Keuper Sandstone.
> The importance of this revolution will be recognized by all persons competent to judge of the subject, whether from a palaeontological or a purely geological point of view.[51]

Huxley's monograph contains little that is new. The intention rather is to write a complete formal account of the research that had taken so much of his time during the previous eighteen years. The general reader might be excused for finding this final publication less interesting than the working papers which preceded it, in as far as Huxley now restricted himself to matters of measurement and analysis – without those outbreaks of speculation. To this *Memoir* he added George Gordon's note on the geology of the region (Appendix III) and Gordon's map, which is reproduced here as Plate 3. An important and impressive feature are the sixteen plates. The separately bound volume, now a rare book, is a monument both to Huxley's tenacity and sense of purpose, and to the methods, standards and professionalism of late Victorian science.

Notwithstanding its monumental qualities, however, one or two further remarks need to be made about this 1877 *Memoir*. In the 'Introductory Remarks' to it, Huxley says:

> But it is a singular fact that, in the course of the nineteen years during which the produce of the quarries about Elgin have been carefully scrutinized, the only remains which have been discovered are those of *Stagonolepis*, *Hyperodapedon*, *Telerpeton* and the jaw with large teeth described in my first paper, and which I then ascribed to *Stagonolepis*; though the subsequent discovery of the true teeth of *Stagonolepis* allows me to correct this error.[52]

In 1859 he had covered his difficulty with the characteristic sentence: 'As the evidence stands, at present, I see no reason to doubt that the jaw belongs to *Stagonolepis*', and to rectify his error he introduced into the 1877 *Memoir* the paragraphs entitled 'A Dentigerous Bone, *Inserte Sedis*', reproducing the fragment of jaw as Figure I of Plate IV. His final sentence reads: 'As it may be convenient that the animal to which this jaw belonged should have a name, I will call it *Dasygnathus longidens*'. This proposal, however, turned out not to be convenient.

By the time A. D. Walker picks up this detail in his 'Triassic Reptiles: Ornithosuchus and the Origins of Carnosaurs',[53] he is in a position to modify both the description and the terminology, since he is aware (in 1963) not of one or two but of eleven specimens. His argument is that what had previously been called *Dasygnathus longidens* should correctly be called *Ornithosuchus longidens (Huxley)*. This name, with 'Huxley' always attached to it, Walker retains throughout his own *Memoir*, irrespective of the names that others have used for the remains of this creature in the interim. On p. 66 what he calls the full synonomy – that is the relationship of genus and species, with a rationalisation of terminology, is set out. Huxley had overlooked the fact that the term *Dasygnathus* had already been appropriated – not a very important error except for someone who, wanting to follow Huxley's work closely, might wonder what had happened to the *Dasygnathus longidens* of *Memoir III*. Anyone asking this question will want to look at Walker's article, both in order to follow his reconstruction of the fourth Elgin reptilian fossil, and to appreciate the detailed work that resulted in the illustrations. Huxley, it needs to be noted, did not return to the study of this fossil after the publication of the 1877 Memoir.

Clearly here is another example of Huxley's work in progress attitude to publication. That he uses the occasion to correct an earlier error is, of course, to be applauded. On the other hand, to give a name to the fossil remains of a reptile of which at that time he had only seen part of a jaw, and on which he didn't do further work, seems cavalier.

A second important section of the 1877 Memoir also deserves comment. This is Section III which he entitled 'The Taxonomy, or Systematic Position, of *Stagonolepis Robertsoni*'. This section which, broadly speaking, is a repetition of the argument of the 1875 article discussed above, has to be compared with the equivalent section of the 1859 paper – the section called 'Affinities of Stagonolepis' – where he says, 'I am at a loss to find an exact parallel for this peculiar combination of characters in any group of recent or fossil *Reptilia*', and then a little later in the same paragraph:

> What little we know, at present, of the laws by which the distribution of life in past time was governed, does not seem to me to enable us to deduce from the existing data any conclusion as to the precise age during which the Elgin Reptile lived. Such combination of characters as it presents would, I apprehend, be in perfect keeping with the known reptilian fauna of any epoch from the Wealdon downwards.[54]

This is at variance with what he had already written in letters. Reasonably enough, perhaps, he preferred to be guarded in public utterances, especially in cases like this one, where he knew further research had to be done. To what extent, then, was Huxley able to refine his position between 1859 and 1877? Anyone grappling with this question on the basis of what Huxley actually wrote in his letters and articles must immediately be struck by his apparent attempt to play down or nullify the special anatomical features of *Stagonolepis*, in order to relate this scarcely known creature to what was securely known. For Huxley, despite stated differences, *Stagonolepis* was still most like a crocodile.

In the 1877 Memoir (as in 1875) he argues that the order *Crocodilia* has three sub-orders, 'distinguishable on purely anatomical grounds': the 'Parasuchia, the Mesosuchia and the Eusuchia', *Stagonolepis* being a genus of the first. He seems never to have corrected this error.

Having established the taxonomy to his own satisfaction, Huxley then writes the following interesting paragraphs, which are here quoted verbatim:

> Hence, if there is any solid historical foundation for the development of evolution, the *Eusuchia* ought to have developed from the *Mesosuchia*, and these from the *Parasuchia*; and if this process of evolution had taken place under such conditions that the skeletons of the *Crocodilia* which have existed during its occurrence have been preserved, geological evidence should prove that the *Parasuchia* have preceded the *Mesosuchia*, and that the *Eusuchia* have succeeded them in order of time.
>
> Now this is exactly what the geological evidence does prove. It is established that these reptiles occur in the following stratigraphical order:
>
> 1. The *Parasuchia*, in the *Trias*, and perhaps earlier
> 2. The *Mesosuchia*, in the Middle Mesozoic rocks from the *Upper Lias* to the *Wealden*
> 3. The *Eusuchia*, in the later Mesozoic rocks and in the Tertiaries from the *Later Cretaceous* to the *Recent Epoch*.
>
> In other words the occurrence of the three divisions of the *Crocodilia* in time, coincides with the order in which they depart from the Lacertilian type, and put on special crocodilian characters: and thus, palaeontological fact is in precise accordance with the needs of the theory of evolution. The evidence in favour of the gradual development of existing from ancient *Crocodilia*, is in fact as cogent, though not so complete, as that by which the origin of the horse from a three-toed ancestor has been demonstrated.
>
> The evolution of *Equus* from its anchitherioid ancestor appears to have nearly occupied the time represented by the Miocene deposits; while the evolution of the Eusuchian from the Parasuchian crocodiles seems to have nearly coincided with the time represented by the deposit of the Jurassic formations.

It is hard to estimate the amount of evolutionary work done in the change of one form into another, in widely different groups in the animal kingdom; but perhaps it will not be very far from the truth, to regard one set of changes as about equal in value to the other.

Admitting the origin of the Horse by a process of gradual modification from some tridactyle or tetradactyle Eocene ungulate mammal, we cannot rest there. This Eocene form must have proceeded from some pentadactyle ungulate mammal, which must have existed before the Tertiary Epoch if it be true, as is commonly believed, that all the known Eocene *Ungulata* possess not even a rudiment of the first digit.

By parity of reasoning, the ancestors of the *Parasuchia* must be sought in the older Triassic deposits, or in formations of still earlier date; and I think it is not difficult to form a probable hypothesis of these ancestral forms, which in Haeckel's nomenclature would be termed *Prosuchia*.[55]

These paragraphs represent Huxley's view of evolution in 1877. Huxley begins in his normal guarded manner. He says: '*If* there is any solid historical foundation to the doctrine of evolution . . .' What would satisfy him would be something observable, something he could regard as factual in his own meaning of the term. He now sees that his sustained study of crocodiles might provide such an assurance. The Elgin reptiles therefore had to be crocodilian. When he began his investigations, he would not have referred to 'evidence in favour of the gradual development of existing from ancient *Crocodilia*', because he was at that time wedded to the notion of persistent types, totally resisting the idea of modification over long periods of time, and even here, having certainly changed his views, he still does not say what he means by gradual development. Having come slowly to the position of accepting evolution (but of course still not in Darwin's sense), he continues to hold away from discussion of its mechanisms. This seems serious, though one could only fairly discuss it in the context of Huxley's career as a whole.

He had been influenced by his visit to Yale in 1876 to see O. M. Marsh, whose work on the evolution of the horse was based on fossil and skeletal finds in the west of the United States, and whose 1874 paper, 'Fossil Horses in America' Huxley must have read in *The American Naturalist*.[56] The principal purpose of this expedition to North America was to deliver the inaugural address at the official opening of the Johns Hopkins University in Baltimore. He also went to see his wife's sister in Tennessee, to Philadelphia, to Buffalo, and to New York City, where he delivered a memorable set of public lectures. Very much in his mind, however, was the problem of whether his conservative opinions as anatomist could accommodate more radical Darwinian ideas about evolution. He would commit himself to evolution only when palaeontology provided the evidence. By the time Huxley made his visit, Marsh was able to demonstrate a sequence of anatomical development using specimens in his possession. Would Huxley have written the paragraphs quoted above had he not been to Yale? Probably not. This would seem to mean that, despite the buzz all around him on the subject of evolution, Huxley had not previously been able to formulate a context or theory into which his palaeontological research would fit. He was not one of the early apostles; he was late to join the converted, and even in 1877 his ideas were still rudimentary. *The Origin of Species* had appeared in 1859 roughly at the time that Huxley became aware of the challenge represented by the discovery of reptilian remains in Moray. Marsh at Yale eventually provided the intellectual or research context that Huxley's work with the chisel needed, so much so that when Huxley took what he had learned in New Haven, and lectured on the evolution of the horse in New York, Marsh remarked drily that it was in the nature of great men to absorb and use other people's ideas. It must have been when he returned to England that Huxley wrote these paragraphs.

Even here, where an advance in his thinking *can* be noted, the attitudes and preferences of the anatomist so prevail in the final paragraph as to leave the earlier, but by no means unimportant phrase 'by a process of gradual modification' unexplained and undiscussed. Similarly, in the passage quoted above, he associates what he calls the *Parasuchia* with the 'Trias and *perhaps earlier*'. It was part of his nature always to hedge his bets, or at least attempt not to leave behind him hostages to fortune.

It will be in the next section that all this will be discussed in more general terms.

Visitors to the Oxford Museum, where Huxley did some of his research on *Crocodilia*, will see the terms 'parasuchia', 'mesosuchia' and 'eusuchia' still in use, but not as applied to the fossil remains of extinct reptiles. Perhaps the crocodile skeleton currently on display is one of those which Huxley saw when he visited Oxford. At all events, as far as the Elgin reptiles were concerned, his preoccupation with crocodiles allowed him to mislead himself. Ironically, he also missed the chance of doing further work on *Ornisuchus* (referred to above), the one late Triassic reptile that might have let him see a link with pterosaurs and dinosaurs.

5

Huxley and his Juliet

In 1869, when Huxley's correspondence with George Gordon was petering out, but well before he had completed his work on the Elgin reptiles, his wife, Henrietta, wrote a letter to a Miss Johnes in which she said:

> I am so very glad that you saw my husband at his own fireside. His writings tell you of his power, but it is in talk with him in his home that you best see his breadth of views, his large heartedness and kindness of nature. I speak as a wife and yet, I venture to think you will say, 'not only so'.[59]

People interested in Huxley have not been able to spend very much time by his fireside, but then he perhaps did not spend enough time there himself. His study at home was a retreat. Henrietta Huxley refers to the standard division in Victorian life between the public image a man necessarily made for himself in the course of a professional career, and the personal life to which only a few would be given access.

If it is polite to accept the image that a man makes for himself, only the indiscretion of not accepting things at face value can permit the biographer the freedom he needs to interpret his subject. A new biography of Huxley is badly needed. In standard accounts of his life, it doesn't help very much to be told about family holidays, or walking expeditions along the Rhine, or in the Alps, with friends – with friends whose conversation was not recorded – if nothing more is being said than that the highly-disciplined, highly organised, highly motivated London scientist allowed himself, for a brief period of time, to be a tourist or a family man. What was going on in his head? That is what we want to be told. When Mrs. Huxley refers to the kindly husband and companion she knew at home, she gently acknowledges that outside the home he was somewhat different. Was the division between the public and the private efficiently, gratifyingly maintained in Huxley's case, or did he have to pay a personal or psychological price for his success?

During the period of his correspondence with George Gordon, Huxley had so fully established himself as a powerful advocate of the social importance of science that it became as natural to defer to his judgement, as it was reasonable for him, in public debate, to do what was expected of him. He was expected to deploy the full force of his rhetoric on behalf of the causes in which he believed. When he spoke, others listened. When he spoke, others *had* to listen. Unfortunately, deferring to his judgement quite often meant delay. But delay might have to be accepted by anyone not willing to cross swords with him.

Sometimes a scheme that others wanted could not proceed, so it seemed, unless it had Huxley's blessing. The launching of the International Scientific Series during the period of Huxley's work on the Elgin reptilian fossils provides an example of this. It was in the early 1870s that the Anglo-American publishing venture, initiated by Appleton in America and taken up by Henry S. King in Britain, had announced its existence by the publication of a first set of important titles, including John Tyndall's *The Forms of Water*.[60] The series was quickly a success. It was nonetheless felt that the contractual arrangements could not be regarded as complete without the active involvement of Huxley, who was a member of the British committee. Why, one wonders, looking back at the complexity of his career, did he bother to give time and energy to a scheme that was already functioning perfectly well without him? And why did he agree to contribute a book to the series when his commitments were already so heavy? At the 1871 meeting of the British Association he found himself agreeing to write a book on ethnology (announced in the earliest advertisements of the series) but he did nothing for the next three years (he was ill) and never contributed the promised book.[61] *The Crayfish* was his later contribution. He was simply too powerful a member of the British three person committee to be ignored. It was therefore he who talked with Henry S. King, the publisher, in 1874 about how the series should be conducted. That the various publishers were already doing what, in 1874, Huxley now said should be done did not detract from the feeling that he had satisfactorily taken the matter in hand. When he wrote to John Tyndall (another member of the British Committee), he kept to himself whatever talk there may have been about his own book. It had been the public figure, the established scientific power broker, who had called on Mr. King. He had not gone to speak about himself.

His Rectorial Address at Aberdeen and his speech at the founding of The Johns Hopkins University are both impressive rhetorically. They are examples of public duties that sustained his influence but kept him away from his work bench. It was expected of him, on occasions such as these, that he would give others direction on how to conduct their lives; it was his duty to do so. If the circumstances of eminence betrayed him into a tone of high didacticism that seems old-fashioned now, many of his contemporaries found it helpful, even inspiring.

Here was an established authority figure telling you how to think and what to think about, endorsing what you considered to be your more advanced ideas and, by personal stamp of approval, liberating you from the unscientific. It would indeed be foolish to dismiss out of hand Huxley's public influence during his own lifetime, or to descend to an ungenerous assessment of what such an energetic, hard-working person decided to do with his time, though it is difficult to avoid the idea that the larger book which eluded him – the one in which the full implications of his research were worked out in a sustained way – was rendered impossible by his restless public life. A person can write many books without writing a good one. Hours spent on the public lecture podium were hours not spent in the laboratory: research reports – occasional papers written at a high standard – were presented to London societies in a continuous intellectual process that often fell short of a fully satisfying treatment of a specialised subject, but which was nonetheless exemplary; but then most scientists have to apportion their time between actually doing scientific work and representing science in a more general way in the public domain. Having chosen to be the spokesman for nineteenth-century science, Huxley was only doing well – with distinction – what others would perforce have had to do somehow.

There were strains and stresses. In a review of Huxley's *Lectures on the Elements of Comparative Anatomy*, William James took Huxley to task for the 'uniformly rude and even malignant tone with which Mr. Huxley speaks of Professor Owen. It is a way he has with all opponents, but especially with Owen, who is so savagely sneered at by name and allusion on every possible occasion. It is a state of things discreditable to science, when, to use the words of an English critic, people go to Professor Huxley's lecture room with somewhat of the same spirit as that with which they flock to a prize fight'.[62] Certainly George Gordon must have been surprised when he refused to look at a specimen that had been sent first to Owen, and more surprised still when Huxley treated it as self-evident that Gordon (like everybody else in the world apparently!) would know that he, Huxley, could not have dealings with Owen. Gordon had after all acted in good faith, ignorant of the tooth and claw of London professional rivalries. In as far as Huxley knew that, it is strange that he replied so sharply to Gordon (see letter dated 29 October 1859) unless he had by then invested his acrimonious relationship with a greater cosmic significance than it could possibly have had. Owen had behaved badly, too. He saw Huxley as an upstart, infidel, rival, enemy, to the extent to which Murchison once had to ask him to stop using Huxley's titles. This, however, was no business of Gordon's, to whom Huxley on this occasion might have behaved in a more relaxed and friendly manner. Unfortunately the pugnacity to which William James referred in his article in the *North American Review* had during the middle part of his career become something of a habit. If he was a bulldog, it was on his own behalf. By comparison, Owen's paragraphs about the *Stagonolepis* in his book on palaeontology seem dignified, factual and restrained.[63]

In a letter to Eper Sargent about an article on 'Spirit Photography' (an attempt to use the camera to prove there were such things as ghosts and apparitions), that had appeared in the *Fortnightly Review*, Wallace says: 'I am informed that Professor Huxley still accounts for the *whole*, by deception and delusion, marvellous acting through a love of the marvellous! No facts or arguments can reach such men as him – nothing but some startling facts occurring in his own family and such are very unlikely'.[64] Though it is easier to sympathise with someone who believes spirits cannot be photographed than with the person who believes they can, this letter, gossip as it is, speaks of reciprocal animosity – and trouble. In a disputed case you had better have Huxley on your side; otherwise he would oppose you. And then!

The strain and stress of battle took its toll on Huxley, affecting the kind of person he became, or was able to be. His health deteriorated. When he did not write the ethnology book he had promised for the International Scientific Series, it was because he was ill. That was through 1872–73, a period in which he also failed to make much progress on the Elgin project. No doubt several factors contributed to Huxley's ill-health and breakdown. One, certainly, was personal and deeply felt. Commiserating with a friend about the death of that friend's son 'whom I should have been very pleased to number among my students', Huxley said:

> Twenty years ago I lost the only son I then possessed and though he was but a child the blow was hard to bear and the memory of it makes my sympathy with your sorrow very genuine.
>
> Patience and hard work pulled me through the evil things of that time, and though I do not think that a man's life can ever be the same after one of those knock-down blows, he may be none the worse fitted to live in this no wise playful world and do a man's work there.[65]

Henry James, the novelist and brother of William James, had met Huxley in the 1870s, had reported him to be a 'very genial, comfortable being – yet with none of the noisy and windy geniality of some folks here', and had described going to the Huxley house on Sunday evenings for general talk 'which he likes to cultivate, for recreation's sake'.[66] They kept up the acquaintance and when, years later, the Huxleys moved to Eastbourne, James had an open invitation to visit and stay with them.

The latter he would not do, partly because it interfered with his work to be away from his own home, and partly because he could not endure other people's domestic rituals. Huxley understood.

> We sympathize so much with your antipathy to 'staying' – which involves being smart and cheerful at breakfast and other inhuman tortures – that you might have safely reckoned upon our effacing ourselves. Anyhow we shall look to you to run in and out as you please – with the certainty of finding people glad to see you – whenever you come to Eastbourne.
>
> Phantoms! My dear friend, I live among them – keeping their damned witches dance at a distance so long as I am strong and well – but at their mercy when [the] liver fails.
>
> A stroll over the downs is the best exorcism – we shall often try it.[67]

An apparently revealing letter of this kind must obviously be interpreted with delicacy, since only a biographical analysis of a far more reaching kind than is being attempted here would allow one to judge for importance of these phantoms that Huxley says he always had to keep at bay. The letter suggests a fear of the irrational, but the basis of that fear would have to be pondered with caution, even if it only took the form of a passionate determination never to be found wrong. Perhaps the type of palaeontology that Huxley practised was his answer to the witches' dance: his solitary labours over the partial remains of extinct fossils, a method of dealing with human problems that it was his habit to repress; the intellectual challenge of the Elgin fossils a satisfactory alternative to the fear of life. Dead reptiles were less threatening than the Euminides. For that matter, they were less threatening than other people.

In the spring of 1873, Darwin, together with a group of friends who were mostly Royal Society fellows and who at the time preferred to remain anonymous, presented Huxley with £2150 (a very large sum of money at that time) to allow him to take a complete rest from work. He had worked himself into the ground. Knowing that the gift had been made in the best possible spirit, from truly disinterested motives and from the respect his colleagues genuinely had for him, Huxley prevailed upon himself to accept, saying however, in his reply to Darwin 'for the first time in my life I have been beaten. I mean morally beaten'. This being 'morally beaten' presumably means not being wholly self-sufficient, having to accept one's limitations (no man is a god) and letting the impersonal public mask slip off to reveal the ordinary, fallible human being beneath it. Huxley said, moreover, that he could not accept the gift without being more frank than usual about his state of health.

> I have for months been without energy and without hope, and haunted by the constant presence of hypochondriacal apprehensions which my reason told me were absurd, but which I could not get rid of.
>
> No, I was breaking down.[68]

Evidently what is meant here is covered by the O.E.D. definition of hypochondria; 'a morbid state of mind, characterised by general depression, melancholy and low spirits, for which there is no real cause'. It will be remembered that Huxley had refused the offer of financial help from the unknown man who had approached him in the Athenaeum. But when the same person died and left a substantial sum to Huxley in his will, Huxley accepted it, saying in his letter to the son that he regretted not remembering the man very well, but that it was probably the same person. Why the lapse of memory? 'Heaven forbid that you should ever be acquainted with the condition that accounts for it'.[69]

In the O.E.D. definition, 'real cause' means an external circumstance that anyone at all would recognise as important, such as a bereavement, a financial crisis, a house burning down or a severe professional disappointment. A psychological cause would, however, be just as real. A manifestation of such a psychological cause might indeed be the dyspepsia to which Leonard Huxley refers, and his condition might have been treated by the kind of remedy his eminent contemporary, the psychiatrist Henry Maudsley tended to recommend – an emetic, beeftea, and exercise – but the word 'dyspepsia' now seems inadequate for a stress related breakdown in health. Furthermore, a person who places a high valuation on being totally in control of his life cannot for all that totally control his relationships with other people, who may have quite different dispositions. When it came to other people, having never to be wrong must have caused a lot of stress. This showed up both in strained relationships and in an overly defensive mode of speech.

In his dealings with George Gordon, his single-minded pursuit of his main objectives – that is, utilising for the purpose of career advancement the knowledge of three extinct fossils the partial remains of which happened to come into his hands before others were aware of them – perhaps somewhat obscured his indebtedness to a man whose devotion to scientific enquiry was at least as disinterested as his own. Gordon had time, as a man of leisure, to take a boat from Lossiemouth to examine the cliffs at Tarbat Ness in the company of like-minded friends. Huxley was not a man of leisure and did not have time for anything other than goal-oriented enquiry. About what is perhaps a slightly inhuman tendency he might well have retorted – 'well, this is the price of being right!' – or anyway, not wrong. But was he right? Did his research stand the test of his own sharpest scrutiny?

The letters show him sometimes failing to respond to Gordon in the right spirit. He shrugs off Gordon's difficulties, wishing only to keep in his possession the specimens he had been sent until his articles had been published, and even then not returning particular fossils that Gordon had specifically asked to have returned. But single-minded ambition and the quirks of temperament are not themselves an adequate explanation of the 'constant presence of hypochondriacal apprehensions'. Did he unconsciously select the type of work that would make him least vulnerable? Was it part of the problem, not only that his palaeontological work – what he actually did when he was by himself – had little immediate human significance (i.e. was disconnected from matters of *living* importance), but also that it lacked the intellectual relevance that it might have had if he had worked out for himself a more plausible life view – that is, a view that was intellectually satisfying because it accommodated his and other people's detailed research. Though he worked hard, Huxley probably never achieved this satisfaction. He tended to isolate each piece of work so that it was invulnerable to attack; it was not his habit to throw discussion open in ways that would leave him unable to dictate its terms. At least as far as his palaeontology was concerned, he preferred to be right on a limited front, even if that meant doing without the conceptual framework that would have given his work greater meaning. He rejected religious explanations, but he lacked any other type of explanation for the phenomena he was studying. He did not at this stage have an explanation for transmutation or for anatomical modification. This left him without the intellectual underpinning that would have made his work more worthwhile and meaningful in his own mind. There is a hint of psychological incoherence here in a man who prized coherence so highly.

Those who do not have orderly disciplined minds, who are carefree and intuitive, or merely disorganised, or uninterested in being so efficient that every minute of every day is well used, sometimes fail to understand the psychological importance of mental orderliness for the person who has the benefit of it, just as a person who has never had a good memory may fail to appreciate how devastating even a minor memory lapse can be for someone whose memory has for years been infallible. Huxley did very much depend upon the great intellectual powers that distinguished him from others: the memory that let him, for example, lecture without notes on great occasions and small; an inexorable sense of the logical; a clear-headed appreciation of the expediency of method; and an ability to make words – the whole apparatus of language – precisely serve his deeply ingrained empiricism. These powers for the most part kept the phantoms at bay. They defined his intellectual territory, making it a world in which it was not dangerous to live. Taking Shakespeare's *Romeo and Juliet* as analogy, he said in one of the more important letters to Gordon, that he rode with neither the Montagues nor the Capulets, implying that the Juliet of his life was *fact*. The Montagues and the Capulets of the world might, if they wished, chase after chimerical ideals, fight, compete, feud, stand upon their honour, hope for better things, cherish their beliefs, defend them, see the personal in political and the political in personal terms – living dangerously, talking wildly – but Huxley preferred to avoid such excesses, searching for an intrinsic orderliness in the world as he found it and much preferring not to be wrong than to be betrayed into untenable speculation, however enticing it might be. As a consciously conceived experiment, his work on the Elgin reptiles eventually reveals both his intellectual strength and, perhaps, his intellectual weakness.

A quick review of how things developed in Moray will serve as a reminder of how cool Huxley remained throughout. Murchison had first looked at the Moray Firth in 1826. Murchison went back with Sedgwick the following year in order to make sense of the geology of both the south and north shores. He returned to London with field notebooks, drawings and fairly detailed sections, including one on a north-south line that took in both Elgin and Lossiemouth. Murchison had the reputation of having a good eye in the field – witness, for example, his work in Wales and, in the Hebrides, his strictures on MacCullough. He had always looked at what he described and he always trusted his own notebooks. A consequence of this was that, although in the late 1820s he had misinterpreted the geological complexities of the Elgin region, too casually assuming uniformity in the Old Red Sandstone as he observed it in various parts of Scotland, he stood by his descriptions for the next thirty years and more, and by doing so greatly influenced lesser lights who were indisposed to challenge his authority without strong evidence. Local studies flourished during this period (1829–59); amateur naturalists assiduously collected the fossil remains of plants and fish they knew to be associated with the Old Red Sandstone, and George Gordon diligently catalogued the flora and fauna of the whole region. All was well until 1859 when Huxley, by arguing and virtually demonstrating that *Stagonolepis* was not a fish but a reptile, threw the previously held geological assumptions into question. Many geologists, including Lyell, travelled to Elgin to see for themselves. Only in 1867, in the fourth edition of *Siluria*, did Murchison concede his error, after which it was left to George Gordon to argue that it was *not impossible* that the rocks in which the reptiles had been found were older than the London experts believed. He wanted Moray to be unique. Huxley knew that it could not be.

Given the talk, controversy, confusion, it is remarkable to notice how calm, unperturbed and patient Huxley remained. Of specimens Gordon sent in 1859, Huxley said: 'alas for Sir Roderick, unquestionable Triassic forms'. On 29 October 1859 he asked: 'Is it true that you still hold to the Devonian [dating] of the Reptiliferous beds?', making clear that he did not. From time to time he expressed, not doubt, but simply the desire that others would catch up with him: 'I wish the age of the famous beds would get itself settled once and for all and then I could bring out my monograph in peace'. In a sense he was lucky. Had he brought out his monograph in peace in 1863, he would have had to do without the new evidence sent from Elgin that allowed him to write 'On a New Specimen of *Telerpeton Elginense*' (1867) and 'On *Hyperodapedon*' (1869), whereas it was in fact these two papers taken together that marked a move away from his previously firmly held convictions about persistent types.

In what way might Huxley's work on the Elgin reptiles most reasonably be assessed? Certainly not in any way that would detract from his achievement. In some uncanny manner he had seized on the chance to examine, analyse, interpret and classify the fragments that he had been sent of three extinct, previously unknown reptiles, *Stagonolepis, Telerpeton* and *Hyperodapedon,* producing a cluster of papers in which he reported minutely on what he had actually done, at the same time using his research as a method of entry into larger questions, such as the meaning of extinction. The quality of this pioneering work was recognised and applauded by his contemporaries, who both appreciated his scholarship and were impressed by his powers of exposition. Furthermore, he achieved his major objective. He had said he would produce a monograph, implying a significant piece of work, and this is exactly what the 1877 *Memoir* is. In the matter of the Elgin reptilian remains, his reputation was by no means negligible.

A number of difficulties and limitations must, however, be noted. The first concerns the dispersal of fossils found in Moray. George Gordon and his friends were impressed by the importance of their region in the context of the new geology they knew to be intellectually important. They wanted this to be recognised and they wanted the fossils to be known by the world at large. These desires remained strong, irrespective of the work being done on identification. When Huxley seemed interested in their finds, they generously sent them to him. When he seemed less than enthusiastic, they sent them elsewhere. Elgin fossils therefore ended up in Marischal College, the Inverness Museum, the British Museum, the Royal Scottish Museum and other places. For example, in 1861, a fossil from Lossiemouth was given to the Glasgow Geological Society.[70] *Stagonolepis* fossils were also sent to a Dr John Anderson in Fife.[71] As far as can be ascertained so long after the event, Huxley's knowledge of this dispersal was imperfect. The letters show that he responded only slowly to Gordon's complaints about the use of the fossils he was sent, having chiefly his own London-based work in mind and only in part appreciating Gordon's motives. Be this as it may, it would seem that he did not get to know as much as he might have done.

A second difficulty concerns the dispersal of fossil fragments on site. Huxley was for the most part willing to rely on Gordon's field work, which in turn depended on Gordon's relations with quarrymen and quarry owners, which he usually reported as being good. The correspondence demonstrates the extent of this reliance. Had Huxley himself spent a few days with Gordon in one or two of the quarries in which reptilian fossils had been found, he might more readily have appreciated that other fragments, though there, had remained uncollected. In his relationship with the earth he was like a lover too busy with other things to make love. He knew, because he was told, that much of the stone quarried was used for building purposes, and that some of it was discarded, and that the amount discarded was sometimes considerable, but he seemed to accept his go-between's assurances that the quarrymen were always on the look-out, and that the rubble of rock left in the worked quarries had been adequately examined. Of course it had not been adequately examined, as later events were to demonstrate. Although strong on the enunciation of general principles, Huxley in this instance declined to do his own field-work. This means that though he was right about the Triassic strata from which the fossils had been extracted, his palaeontology was less thorough than it might have been.

A third difficulty concerns what he did in his laboratory. Exactly how he worked is difficult to ascertain but, if one is to judge by his correspondence and published papers, his main effort seems to have been the one of liberating actual fossils from the rock or matrix. The chisel was his instrument. There is no mention of a drill. Sometimes Gordon sent him reconstructions made in plaster from the mould or counterpart. Sometimes he made these himself. The main purpose of the preparation was the orthodox one of freeing the physical evidence, which he then measured with scrupulous care for the sake of skeletal extrapolation. He could only work with what he had been sent, though, and Gordon's selection of 'remarkable' specimens eliminated parts of the evidence that a palaeontologist ought to have wanted to examine.

When Huxley turned to other subjects of enquiry and other concerns, there remained much to be done before the fossil wealth of Moray was fully appreciated and before many of the problems which Huxley had identified in his pioneering work had been properly treated. With the appointment of a new generation of geologists to university chairs, the direct involvement of Archibald Geikie as Director of the Geological Survey in Scotland, the actual survey work itself in the North-East, and the fresh impetus provided by museums in Edinburgh and Glasgow – all these strengthened the study of both geology and palaeontology in

Scotland, much more work of professional quality being done by people who lived there. In England, a young colleague of Huxley's, Edwin Tully Newton, took up the subject of the Elgin fossils, publishing two important memoirs in the *Transactions* of the Royal Society.

At a later date, in 1960, A. D. Walker, then in the Department of Geology at King's College, Newcastle-upon-Tyne, published two articles or memoirs which were intended as part of a set in which all the problems of the Elgin Sandstones would be discussed. These two articles were: 'Triassic Reptiles from the Elgin Area: *Stagonolepis, Dasygnathus* and their Allies' (*Philosophical Transactions* Volume 244, pp. 103–204, 31 August 1961) and 'Triassic Reptiles from the Elgin Area: *Ornithosuchus* and the Origin of Carnosaurs' (*Philosophical Transactions* Volume 248, pp. 53–134, 26 November 1964).

In order to write these articles, Dr Walker examined and assessed all the specimens by that time known (as opposed to the limited number Huxley knew) and at the same time considered carefully the work of all previous palaeontologists who had interested themselves in the subject. In the *Stagonolepis* article, Walker notes that the specimens in the Elgin Museum carry two sets of identifying numbers (see Plate 5). One of these, in red paint, is said to be Huxley's original numbering and the one used in his 1877 Memoir. The other specimens, with black numerals on white labels, were according to Walker not necessarily unknown to Huxley, though unmarked by him. Walker was well aware of the difficult circumstances under which Huxley had worked. Whereas he could say on his own account, for example, that the 'major portion of the material of *Stagonolepis* in the British Museum can be fitted together thus proving the association of one individual', he recognised that Huxley had never been in the position to attempt this. Likewise when he says that his description of the *Stagonolepis* skull will be based on specimens in Marischal College (then) and the British Museum, he knows that Huxley had seen neither of them.

It is not necessary here to follow Walker through the disciplined, elaborate account in which he makes sense of *all* the surviving fossil fragments of *Stagonolepis*, but it may be useful to take note of what he says about Huxley, because by that means one is able to appreciate more clearly Huxley's awkward initial predicament. For example when Walker describes his 'Material and Methods' at the beginning of his 1961 Memoir he says:

> Huxley remarked on the incompleteness of many bones, but this is chiefly due to the fact that all the pieces were not collected. On the other hand the material he described certainly includes many damaged specimens: in the case of Lossiemouth it is possible that these came from a slightly different horizon from those of later

> finds, containing somewhat dissociated or partial skeletons. Material presumably collected later from Lossiemouth, including that presented to the British Museum by the Rev. George Gordon, is often better preserved and includes various articulated skeletons of varying degrees of completeness.[72]

It has already been mentioned that Huxley was never in a position to know whether or not all the pieces had been collected. Though he said that his Juliet was Fact, his facts could only be those of which he happened to be aware.

Walker came to the conclusion that the *Stagonolepis* remains provided evidence of reptiles of two different sizes, to establish which thesis he had, of course, to be completely confident about the relationship of the remains. In this connection he says:

> In only one case has there proved to be a mingling of elements of two individuals of *Stagonolepis*, but even here the intruding ilium is on the back of the slab displaying the other bones and some inches of matrix separates them. Acceptance of this as a true association mislead Huxley in his estimate of the proportions of the pelvis.[73]

Likewise, in an important section on the *Stagonolepis* skull, Walker says:

> E.M.38R (the number code for a specimen in the Elgin Museum) was wrongly interpreted by Huxley; it shows in actual fact the inner aspect of the bones which surround the left external naris. The specimen provides an important key to the understanding of the relationships of *Stagonolepis*, since when correctly identified the large naris and premaxillary tip are strikingly similar to those of *Aetosaurus ferratus*.[74]

Walker is gently challenging Huxley's crocodilian thesis.

Because the present purpose is simply to draw attention to the difficulty of the tasks Huxley set himself, rather than to establish in absolute terms the extent to which he was right or wrong from the point of view of modern palaeontologists, other references to Walker's Royal Society paper will be placed in the notes to the letter texts where appropriate. It would in any case be difficult to arrive at an unassailable conclusion. Indeed, from the point of view of the present writer (a literary man), towards the top of the list of words most preferred by geologists and palaeontologists are the key terms 'apparently' and 'probably'. In J. P. B. Lowell's article entitled 'Moray Firth Basin and Northern North Sea', he says:

> The older fossils come from the aeolian Cuttie's Hillock (now Hopeman) Sandstone west of Elgin. The remains are nearly all referable to two groups, the dicynodonts or anomodonts – and the pereisaurians. They are of late Permian to Early Triassic age, more probably the latter – The unfossilerious Burghead Beds are faulted against these aeolian Sandstones, and are thought to be younger. The aeolian Lossiemouth Sandstone contains a reptilian fauna of probable late Triassic age.[75]

Passages of this kind, troublesome as they are, at least have the effect of making Huxley's caution perfectly legitimate.

Similarly, in A. L. Mason's *The Growth and Structure of Scotland* there is the statement:

> A similar marginal relationship is thought to have existed in the Moray Firth area, but here the marginal sediments consist dominantly of aeolian deposits, notably sand dunes, which have yielded both very early and late reptile faunas. Here early and late aeolian rocks are apparently separated by waterlain sandstones carrying abundant well rounded pebbles.[76]

Apparently? It might be worth having a look.

If these tempered, cautious statements, hanging from 'probably' and 'apparently', and other phrases such as 'is thought to have existed', are legitimate, as they no doubt are, one would be all the more disinclined to fault Huxley for the pioneering work he was able to do on the basis of less evidence than now exists. Better, surely, to give him credit for initiating a line of research that has taxed the wits of other able investigators during the hundred years and more since his 1877 Memoir was published.

By bull-dog standards he, too, was extremely cautious. As Mario di Gregorio has pointed out in *T. H. Huxley's Place in Natural Science*, Huxley was slow to shift his position from a preoccupation with persistent types to an interest in the question of descent. When he made this shift, at some point between 1866 and his 1874 visit to the United States, and before he published his 1877 Memoir on *Stagonolepis*, he was at pains to establish the relationship of *Stagonolepis* to the sub-order *Crocodilia*, of which he had for many years been making a careful study. In this regard, however, di Gregorio remarks: 'it is odd that the question of the mechanism of descent is nowhere addressed – or even raised – in Huxley's application of the notion [of the transmutation of animal forms]'.[77] This is an important comment since it identifies one of the boundaries of Huxley's thought. He could not, of course, formulate a general theory of descent on the basis of single specimens, least of all when his best efforts had been devoted first to reconstructing them, and then to establishing their taxonomic placement (though not accurately). But why did he not want to formulate a general theory? And why was he disinclined to do the work that would put him in a position to formulate one?

In his book *Wonderful Life*, which is about an even more remarkable find than the Elgin reptilian fossils (that is, the Burgess Shale in British Columbia) Stephen Jay Gould engages in a conscientious assessment of the scientific career of Charles Doolittle Walcott, a scientific administrator whom Gould entirely respects and the person most responsible for the early work on the Burgess Shale. Why did Walcott not understand the significance of the site he himself had discovered, and of the fossils he himself had collected and preserved in the Smithsonian? In answering these questions, Gould says:

> He [Walcott] never found even the entering wedge of time to examine, ponder, ruminate, observe again, obsess, reconsider and eventually publish – the essential (and incomprehensible) ingredients of a proper study of these complex and precious fossils.[78]

These remarks would appear to apply equally to Huxley who, like Walcott, felt it his duty to be a spokesman for science, and who therefore left himself insufficient time either for sustained work in a particular field, or for thinking out the implications of the research he and his contemporaries were doing. As indicated earlier, he seems to have relied on others for the conceptual framework within which research of this kind might be regarded, his ideas coming mostly from his reading. So he was a palaeontologist with a difference. His claim that fact was his Juliet failed to take into account the misunderstandings, cross purposes and tragic conclusion of *Romeo and Juliet*. The devotion was real but the love unconsummated. Perhaps a different lover would have done more for him than was within the innocent Juliet's powers. It no more helps to be star-crossed in scientific enquiry than in any other human enterprise.

What conclusions, if any, may be drawn from this story of Huxley and the Elgin reptiles?

Huxley sent Gordon a copy of *Memoir III of the Geological Survey*, which Gordon acknowledged, though it seems to have been sent to him without any accompanying communication. Huxley had no further interest in Gordon's activities in the field; their working relationship effectively ended in 1877. But George Gordon's interest in the Moray District nonetheless continued until his death, so that he could be said to have participated in the two main phases of work that characterised nineteenth-century geology in Scotland; the first, the explorations that were begun by MacCullough and Miller, and continued by Murchison and Sedgwick; the second, the more systematic enquiries conducted under the auspices of the Geological Survey of Scotland by Archibald Geikie and his associates. Gordon first communicated with Geikie in 1876, though they did not meet until the following year, the year of Huxley's *Memoir*. By that time, Geikie's survey team – his 'three fine fellows', as Gordon called them – was at work in the North-East, and it will come as no surprise that Gordon assisted them in the field in every way possible, and also corresponded with them exactly as he had done with Huxley, albeit on less sensational subjects. As the work of the Geological Survey became more precise and authoritative, he necessarily saw out and abandoned his cherished ideas about the Old Red Sandstone, it being to his credit that, having put up a strong defence under the influence of Hugh Miller and R. I. Murchison, he moved with the times and, though ageing, participated actively in the research of a new generation of geologists and palaeontologists.

Gordon's post-Huxley correspondence is by no means negligible, stretching as it does from the interesting exchange with John Judd in the mid-1870s to the letters he wrote to and received from E. T. Newton in the early nineties, when Newton was taking up again the study of the Elgin reptilian remains. That Newton had more specimens to work with than Huxley had had was largely due to the continuous field work that has already been noted. So far, only small portions of the correspondence that represented Gordon's life work have been subjected to intensive study and one therefore hopes that the Elgin archive will one day be honoured by a more comprehensive publication that does it full justice.

Huxley may not have appreciated the extent to which he had become a figure of fun in the North-East. Perhaps he was blinded by a sense of metropolitan superiority to the fact that he was dealing with intelligent, independent people who were as interested in his subject as he was, though they did not publish memoirs in London. Instead they built a Museum for the edification of their own people. In a letter dated 13 May 1893 E. T. Newton gave Gordon a report of a London lecture:

> At a lecture I gave on the Elgin reptiles I showed the sketch you sent me made by Mr. Joass. Murchison and Agassiz fighting with slabs of stone & Huxley reclining on his reptiles. It was highly appreciated as was also the portrait of Dr. G. Gordon which I took the liberty of showing on the screen.[79]

There are other cartoons by Joass in the Elgin archive, one showing Huxley on a beach in frock coat and top hat being startled by the emergence from a cave of a fearsome, crocodilian monster. The difference is that Edwin Newton went to the Moray Firth where, sharing their boat trips and their jokes and their local talk, he gained the confidence of the people with whom he worked, whereas Huxley did not. But who would take with anything but good humour the blind spots of the Metropolitan sensibility?

Huxley was only fifty-two when the *Stagonolepis* memoir was published, but he seems not to have kept his research going after that, except to the extent of writing once again about the *Hyperodapedon*, as has been mentioned above. He was by this time even more the great public figure; the temptation to pontificate was enormous. One can therefore look back over his forays into palaeontological research, seeing clearly enough that his contributions during the eighteen-year period being discussed here were significant ones. He had carried forward the study of the subject in ways which the next generation of scientists found helpful. And he had done pioneering work, which was not only useful to the specialist but also brought the methods and purposes of scientific enquiry to the attention of the public – an enlarged public, more aware in 1877 than it had been in 1859 of the intellectual, as well as the technological importance of science.

From a personal point of view, however, he may have missed a grand opportunity by being insufficiently reflective, by being too preoccupied with what he called 'fact' (a term not satisfactorily defined by him) and from deriving too much of a feeling of security from the quite simple type of empiricism within whose boundaries he was prepared to stay. Furthermore the rhetoric that impressed, and still impresses, the humanist was, for a scientist, too much of a liability, particularly when it took the form of those sets of negatives that were designed to keep opponents at bay, or had that effect. Huxley found it too easy to say that the evidence did not exist to support particular groupings of ideas, and because in this he was so often right, he impressed people and, in an important sense, taught them to think, or sharpened their logic. Photographing spirits indeed! But these – could one say? – lapses into negative rhetoric somewhat impeded the quest for new evidence. Someone else would have to find the new evidence. In other words, the deeply puzzling psychological caution represented by these rhetorical negatives does seem to define the type of scientist he was. This is true, certainly, of his work on the reptilian fossils in Moray.

Stephen Jay Gould well expressed an aspect of this when he reviewed Mario di Gregorio's *T. H. Huxley's Place in Natural Science* in the *Times Literary Supplement*. Gould wrote:

> More central, and here I praise di Gregorio for an important insight, Huxley, for all his brilliant words on the methods of science, never really grasped the essential contrast between stereotypes of procedure

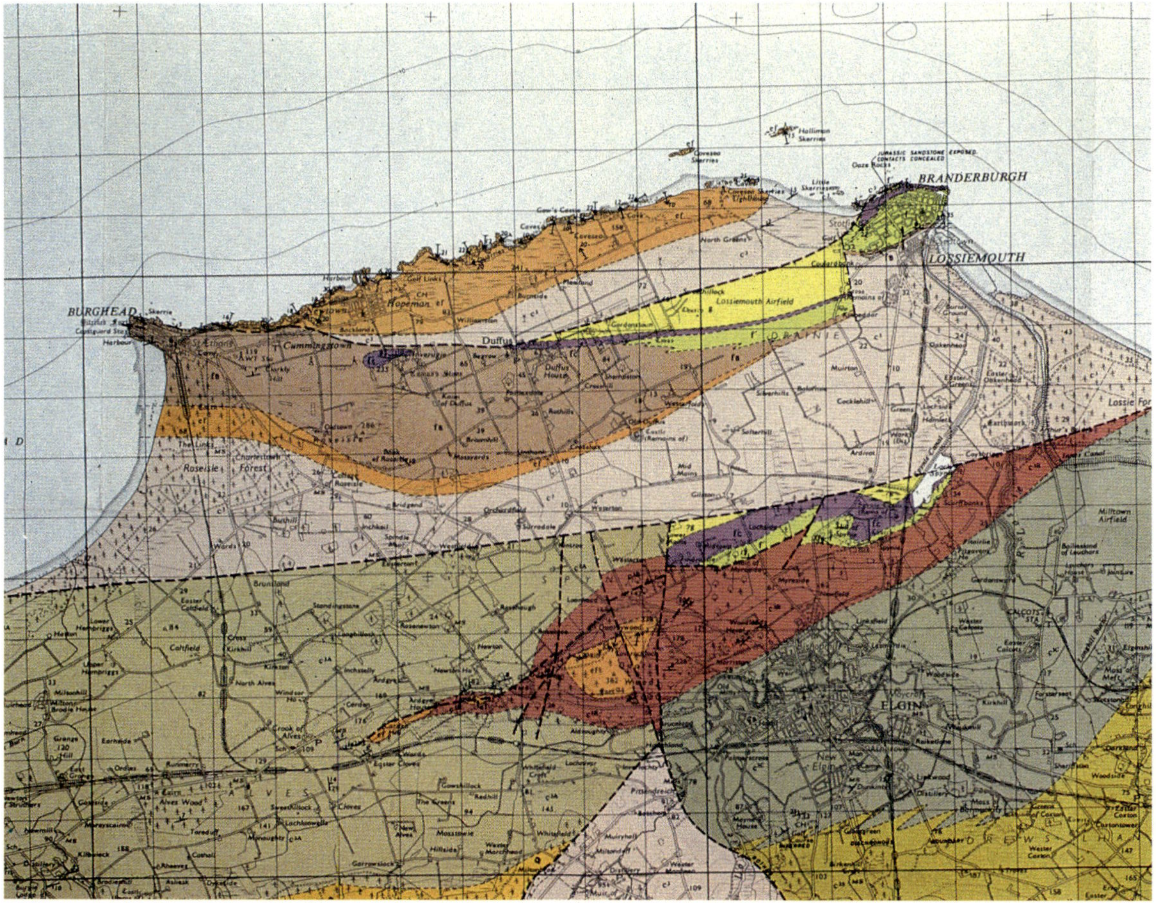

Plate 1. A modern geological map of the Elgin District. The colour coding, showing different rock types, indicates the stratigraphical complexity Gordon and contemporary visiting geologists failed to observe from the surface. (Reproduced by kind permission of *The Geological Survey*).

Plate 2. Section of the 1876 Ordnance Survey Sheet 95, as revised in 1896, showing the Elgin area. This map does not differ in

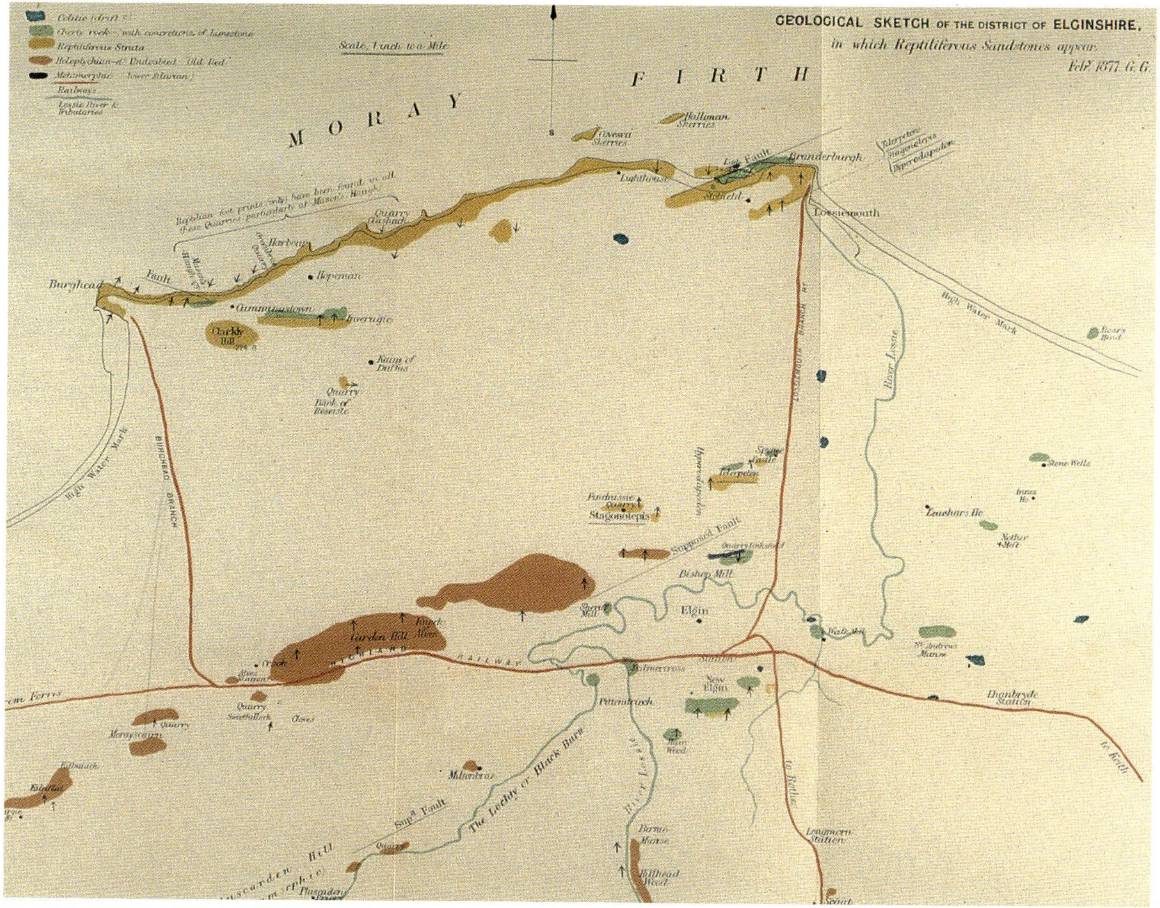

Plate 3. Geological sketch of the District of Elginshire, by George Gordon. This map was drawn at Huxley's request to indicate the location of the fossil finds referred to in Huxley's Monographs.

Plate 4. T. H. Huxley in 1857.

Plate 5. (a) Teeth and (b) scuter of *Stagonolepsis*. These were among the first of the Elgin fossils examined by Huxley, and carry the identification marker referred to in the correspondence. The fossils are now in the Elgin Museum (ref. 1978.563.5a and 1978.562.6 respectively).

Plate 6. The type specimen of *Hyperodapeilon gordoni*, from the Elgin Museum (ref. 1978.56b).

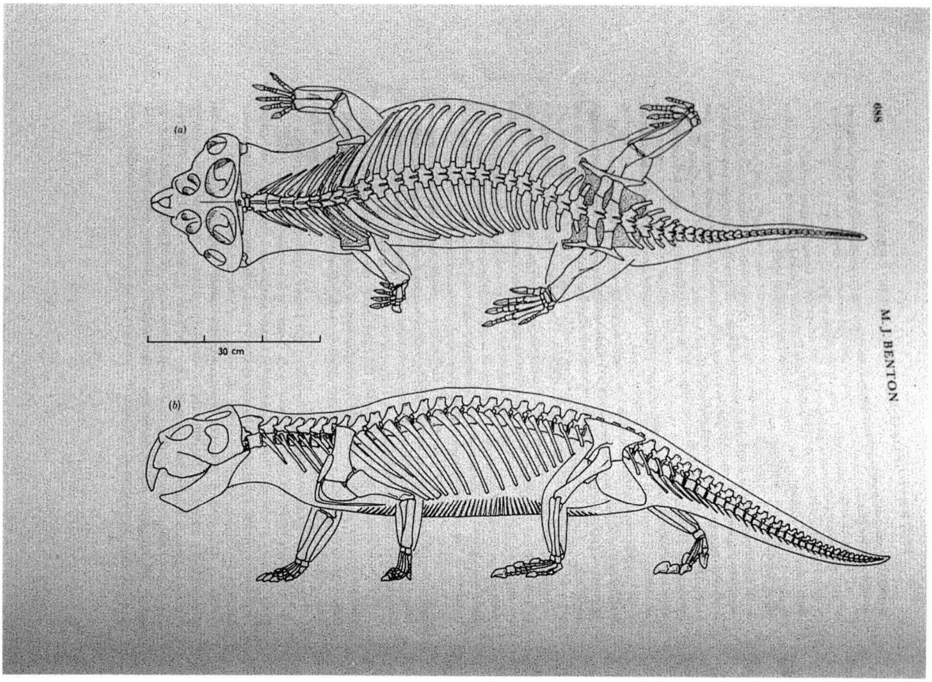

Plate 7. Skeletal reconstruction of *Hyperodapeilon gordoni*, by Dr. Michael Benton. (Reproduced by kind permission from *The Triassic Reptile* Hyperodapeilon *from Elgin*, Benton, 1983).

Plate 8. The cliffs at Coversea, referred to in the early correspondence.

> in conventional physical sciences and the different methods that the complex historical sciences must use. Huxley, committed to strict empiricism and the crucial experiment, sought a convincing experimental proof for evolution. But history is a web of irreducibly complex and unrepeatable events, each one so intricate and so full of ambiguity that one cannot hope for a clean single experiment as proof of any particular theory of mechanism. Huxley wrote endlessly about method and missed this distinction between what his profession did and what the stereotype of science proclaimed as universal about procedure. Darwin, who chose to say so little about 'philosophy', but who understood so deeply with his distinctive uncanny intuition, recognized the difference. Darwin knew that the search for a convincing experiment was a chimera. Natural selection must seek another source of support – confluence of inductions if you will. We will accept it provisionally if it unifies and makes sense of a large set of phenomena, heretofore disjointed and unexplained – 'this hypothesis may be tested – by trying whether it explains several large and independent classes of facts; such as the geological succession of organic beings, their distribution in past and present times, and their mutual affinities and homologies. If the principle of natural selection does explain these and other large bodies of facts, it ought to be received' (Darwin, 1868). Ultimately then, and for good reason, the nineteenth century belongs to Darwin. Yet how sorry and shabby a time it might have been without Huxley.[80]

These are convincing and, as well, generous words. In searching energetically for his proper role as scientist, for the way in which he could make a useful, indeed significant contributions to the profession in which he so strongly believed, Huxley undoubtedly put heart and soul into his work, teaching himself and teaching others how to conduct a certain type of research, and how to think. He could only go so far; others would have to complete the enquiries he began. There is hardly anyone, whatever the field of study, whose labours, however earnest and disciplined, are not inhibited by some form of psychological restraint, some limitation in the person, some set of personal defining factors that allow that person to go so far, but no further. The psychological imperatives in Huxley's case had to do with the fear of being put down by another human being, and the peculiar blindness that goes with that condition. At the point of contact between one person and another Huxley more wanted to prevail than to submit, more wanted to be right than to understand why the other also thought he was right and, obviously but painfully, more needed not to be wrong in the eyes of those he respected or feared than to enmesh himself in the warm, merely human confusions of collaboration. Because this was his bent, it seems friendlier to accept the fallibilities and limitations of a scientist who achieved so much by dint of the ever forwards progress of his activities, than to search out, as though to judge by, the absolute, world-altering truths and stunning research breakthroughs that, in Huxley's case, may not be there to be found. But what might he not have done had he not been so wholly repressed, as his language shows him to have been?

There is an especial charm to the later photograph that is used in the present book as frontispiece. Here is the older Huxley, the established public figure dressed appropriately in a well-tailored cut-away, and characteristically holding a book, perhaps even the well-used copy of *Paradise Lost* that was the only text he referred to in his New York lectures. (He could associate with Adam's self-consciousness and the loss of God). If the hint of carelessness evinced in the button that needs to be sewn on more securely and the tie that needs to be adjusted speaks of a raffish bohemianism, it is a bohemianism more than counterbalanced by the firm set of the jaw and the unflinching, steadfast gaze of the eyes. Here is Victorian high-seriousness exemplified. If the world is all in all a pretty bad place, you had best get yourself a tailor and do something in the public forum to improve matters, if they cannot be mended. Science was the means by which this utilitarian accommodation could be attempted. And for the purposes of non-religious intellectual adjustment, palaeontology was one of the fields in which Huxley found he could most comfortably work.

Notes to Part One

ABBREVIATIONS USED IN THE NOTES

LL Leonard Huxley, *Life and Letters of Thomas Henry Huxley* (London: Macmillan & Co. 1903; in three volumes).

M *The Scientific Memoirs of Thomas Henry Huxley* (ed. Michael Foster & E. Ray Lankester), London: Macmillan & Co., 1898 (5 volumes).

1. See Ruth Barton, '"An Influential Set of Chaps": The X-Club and Royal Society Politics 1864–1885 ', *The British Journal for the History of Science* 23 (March 1990), 53–81.
2. ALS 9 November 1851; Huxley to John Tyndall (Imperial College Archive). Compare this with his letter to Geikie, dated 19 October 1862, where he says the meetings at Cambridge had proceeded 'with just fighting enough to prevent one from getting "blue mowldy for want of a bating"'. (University of Edinburgh Library)
3. ALS Huxley to Knowler, 19 December 1870: American Philosophical Society.
4. University of Oklahoma: Special Collections, 'Evolution' (B4–F21).
5. For an excellent listing of Huxley's papers see Mario di Gregorio's *T. H. Huxley's Place in Natural Science* (New Haven and London: Yale University Press, 1984).
6. ALS March 1876 and 1 April 1876: William R. Perkins Library, Duke University.
7. J. A. Secord reviewing Robert Stafford's *Scientific Empire: Sir Roderick Murchison, Scientific Exploration and Victorian Imperialism, Times Higher Educational Supplement*, 4 May 1990, p. 25
8. ALS 14 May 1863; Huxley to Archibald Geikie (University of Edinburgh Library).
9. ALS 25 March 1875; Huxley to Professor Blackie (National Library of Scotland: 2631.f.305).
10. Roy Porter, *The Making of Geology: Earth Sciences in Britain 1666–1815* (Cambridge: Cambridge University Press, 1977).
11. *M* II, pp. 513–14.
12. *M* IV, pp. 656–57.
13. Anderson's *Guide to the Highlands and Islands of Scotland* (1851) footnote to pp. 344–49.
14. Mario di Gregorio, *Huxley's Place in Natural Science*, p. 68.
15. See Dr Malcomson's letter to George Gordon dated 12 June 1835 (Elgin Museum DDW/839/8) which refers to the fact that the fossil finds of W. A. Stables, John Martin and Malcomson were then sent to Louis Agassiz. This was the beginning of the story. Huxley later had distant dealings with Agassiz. There is a letter written at the time when he

was beginning to work on the Elgin fossils in which he hopes that Agassiz will regard him 'neither among the cantankerous, not even the bumptious' (ALS Huxley to Agassiz 11 July 1959, Houghton Library), a letter which does not suggest a close working relationship. Much later Huxley told Agassiz' son that he had never known his father well, though he recollected drawing room meetings in London. This letter to the son does scant justice to the extent of Huxley's indebtedness to such an influential figure.

16. Georges Cuvier, *Essay on the Theory of the Earth* (trans. Robert Kemp), 1812. See also Cuvier's *Discours sur les révolutions de la surface du globe, et sur les changements qu'elles ont produit dans le règne animal* (1826) and also his *Anatomie comparé* (1850–56).
17. Daniel Defoe, *A Tour through the Whole Island of Great Britain*.
18. Patrick Duff, *Sketch of the Geology of Moray*, p. 55.
19. Elgin *Courant & Courier*, 15 December 1882.
20. Ibid.
21. Roderick Impey Murchison was Huxley's senior by many years and, as Director of the Geological Survey, his administrative superior. He had already written about the territory in which the reptiles were now being found. Understandably Huxley was reluctant to cross swords with so well-established a figure. The two men were different from each other in background, education and attitude. It seems correct to say that the full course of Huxley's career cannot be described until his somewhat obsequious relationship with Murchison has been worked out. Had Huxley felt himself to be Murchison's equal he would surely have been more forthright.
22. *M* II, p. 95.
23. *M* II, p. 99.
24. *M* II, p. 107.
25. *M* II, pp. 110–11.
26. *M* II, p. 117.
27. *M* II, pp. 117–18.
28. *LL* I,, pp. 251–52.
29. ALS Symonds to Lyell, dated 2 January 1863 (University of Edinburgh Library).
30. *M* III, p. 205.
31. *M* III, p. 212
32. *M* III, pp. 212–13.
33. 'On Hyperodapedon' *Quarterly Journal of the Geological Society*, 25 (1869): 138–52. *M* III, pp. 374–90.
34. *M* III, p. 376.
35. *M* IV, p. 644.
36. *M* III, pp. 283–4.
37. *M* III, p. 386.
38. *M* III, p. 387.
39. David Lambert, *The Cambridge Field Guide to Prehistoric Life* (Cambridge: Cambridge University Press, 1985), p. 115.
40. *M* III, p. 390.
41. *M* III, p. 390.

42. M. J. Benton, 'The Triassic reptile *Hyperodapedon* from Elgin: functional morphology and relationships', *Philosophical Transactions*, Volume 302, Number 1112, p. 601.
43. Ibid. p. 614.
44. *M* IV, p. 67.
45. *M* IV, p. 70.
46. *M* IV, p. 77–78.
47. *M* IV, p. 78.
48. Mario di Gregorio, *T. H. Huxley's Place in Natural Science*, p. 96.
49. Ibid. 96.
50. *M* IV, pp. 82–3.
51. *M* IV, p. 83.
52. 'The Crocodilian Remains Found in the Elgin Sandstones, with Remarks on the Ichnites of Cummingstone', *Memoirs of the Geological Survey of the United Kingdom*, Monograph III, 1877.
53. Ibid. p. 3.
54. Ibid. p. 8.
55. A. D. Walker, 'Triassic Reptiles of the Elgin Region: *Ornithosuchus* and the Origins of Carnosaurs'.
56. Monograph III, p. 113.
57. Monograph III, pp. 42–43.
58. Marsh, O. M., 'Fossil Horses in America', *American Naturalist*, 8, (1874): 288–94.
59. ALS Mrs. Huxley to Miss Johnes, 16 March 1869 (National Library of Wales: Dolauccthi L 7066).
Compare this with Leslie Stephen's earlier remark in a letter dated 29 December 1866 to Minny Thackeray where he says Huxley 'is really a most agreeable man considering that he is a scientific swell', and in the same letter, 'after dinner Huxley joined our smoking party and we had a good talk – denouncing God, Eyre and the British aristocracy by the hour'. (Leslie Stephen's papers, Perkins Library, Duke University).
60. John Tyndall, *The Forms of Water*, London: Henry S. King, 1872.
61. The International Scientific Series was initiated by the American publisher, Appleton, and involved publishers in five other countries: Britain, France, Italy, Germany and Russia. The publisher in each country was to have the advice of a three person academic committee. Huxley was a member of the British Committee, with John Tyndall and Herbert Spencer.
62. *The North American Review*, December 1884.
63. See R. Owen, *Palaeontology, or A Systematic Summary of Extinct Animals and their Geological Relations*, Adam & Charles Black, 1860, pp. 254–7.
64. ALS Wallace to Eper Sargent, dated 9 July 1874 (Boston Public Library: MS Acc 434).
The correspondence refers to an article in the *Fortnightly Review* entitled 'Spirit Photography'.
65. ALS dated Sept 1860 from 6 Barn Park Terrace, Teignmouth (University of Kentucky Library).
66. Leon Edel *Henry James: The Conquest of London 1870–1881*.

67. ALS Huxley to Henry James, dated 3 November 1894 (Houghton Library).
68. ALS Huxley to Darwin, dated 24 April 1873 (The Royal Institution).
69. ALS Huxley to an unnamed correspondent: William R. Perkins Library, Duke University
70. ALS from John Anderson of Fife (Elgin Museum DDW 71/859/3) acknowledging the receipt of *Stagonolepis* specimens.
71. Walker, p. 106.
72. Ibid. p. 110.
73. Ibid. p. 114.
74. J. P. B. Lowell, *Geology of Scotland* (Edinburgh: Scottish Universities Press, 2nd Edition, 1983), pp. 334–42.
75. A. L. Mason, *The Growth and Structure of Scotland*, pp. 20–21.
76. Mario di Gregorio, *T. H. Huxley's Place in Natural Science.*
77. Stephen Jay Gould, *Wonderful Life. The Burgess Shale and the Nature of History* (New York: W. W. Norton & Co. 1989).
78. DDW 71/893/11
79. *The Times Literary Supplement*, 18 January 1985, p. 70.

Part Two

The Correspondence

Part Two consists of 32 letters from T. H. Huxley to George Gordon, 27 letters from Gordon to Huxley, 1 letter from Gordon to Sir Henry James, and 1 letter from Huxley to Patrick Duff.

All but two of the Gordon letters are autograph drafts. Only in two cases have the letters actually received by Huxley from Gordon been preserved, although it can be seen from the letter of 10 August 1866 that the difference between the draft that Gordon retained and the letter he sent was not necessarily great. Gordon's drafts take different forms, ranging from verbatim copies of letters as sent, to excerpts of parts of letters he evidently wished to remember accurately, and what are so obviously rough first attempts that the letter as sent may have been distinctly different. The letter from Gordon to Sir Henry James is included only because it relates directly to a sequence of letters between Huxley and Gordon. As indicated in the Preface, the bulk of the correspondence is now in the Elgin Museum. All the Huxley letters are previously unpublished autograph originals.

To preserve the integrity of this presentation of a scientific correspondence between two people urgently interested in the same thing, letters by Huxley and Gordon to other correspondents on the subject of the Elgin reptiles have not been included, though they are occasionally referred to in Part One and in the notes. While Huxley, because of his clearly defined objectives, wrote to only one or two other people to check details, Gordon's correspondence was vast. From the Manse at Birnie he wrote to almost anyone who could be expected to show an interest in the natural history of Moray, including most contemporary Scottish geologists. These excursions of his would be an important part of any life of Gordon, or any history of geology in Scotland, but they are not directly relevant here. Here the purpose is to focus on that collaboration between Huxley and Gordon that resulted in a set of major Huxley publications, at the same time advancing the study of palaeontology itself.

At the same time as Gordon was writing to Huxley as palaeontologist, he was also writing, as geologist, to Roderick Impey Murchison. The interlocking Murchison letters will be published separately.

There are several reasons for this policy. Even though Murchison was among the first to recognize the reptilian nature of the Elgin fossils, he himself was not a practising palaeontologist. He in effect passed the work on to Huxley. By contrast, while Huxley demonstrated little general interest in the geology of Scotland, Murchison travelled concerned himself actively with the geological enquiries that interested Murchison, it would have been more difficult in this book to distinguish between the geology of the one and the palaeontology of the other. Gordon himself did not see this clearly at first, when both men were simply London *savants* of known eminence. Consequently, he went on writing for a while to Murchison about matters Murchison had delegated to Huxley. In what follows, the details of how Murchison and Huxley decided to proceed are dealt with in the notes. Though there is no record of Murchison's and Huxley's talk about who should do what, the chronology of Huxley's palaeontological enquiries can be worked out and demonstrated clearly, while during the same period Murchison was deeply involved in the affairs of the Royal Geographical Society, having already produced his major geological work.

Huxley by no means kept all the letters he received, though he kept a great number. The Huxley–Gordon correspondence is typical of Huxley's practice in general or, rather, of his lack of any system for filing and preserving letters received from any particular correspondent. There are therefore gaps in the exchange of letters that is reproduced here. In his very first communication Huxley says: 'I have just been examining your last consignment', though this is not preceded by a known letter from Gordon, that is, not a known letter from Gordon to Huxley. Other letters, as that of 15 November 1858, must have been preceded by a Gordon letter that has not survived. Despite these gaps the correspondence does have a shape – a shape determined not by Huxley's research but by his publication of articles or memoirs. The relatively vigorous correspondence of 1858 and 1859 leads up to his first publication on *Stagonolepis*. This is quickly followed by a paper on the *Hyperodapedon*. In 1859 Huxley met Gordon at the meetings of the British Association in Aberdeen. In the same year, Gordon was awarded an honorary degree by Marischal College. But these events do not seem to have resulted in a strengthening of the relationship. No letters have survived from 1860 and 1861, and the correspondence does not resume until August 1862, when there are further exchanges which lead up to Huxley's paper on the *Telerpeton* in 1866. At this point, he had said all he had to say about the Elgin fossils, but he still had to produce the long overdue monograph, so the correspondence continued until it was published, at which point Huxley lost interest both in George Gordon and in the new research that younger people were beginning to do on the same material. Of course all his correspondence had been interrupted by the illness which put him out of action through much of 1870 to 1872.

In short, despite the gaps caused by missing letters, the correspondence is ancillary to Huxley's publications in the same way as Murchison's field notebooks, for example, are ancillary to his.

A thorough survey of libraries in Britain, the United States and elsewhere seems to show that there are no other letters from Gordon to Huxley, or Huxley to Gordon, apart from those included here, though of course the author would be glad to hear of any that slipped through his net. All the libraries in the American Association of Research Libraries were consulted, as were other major libraries in North America. There are many uncollected Huxley letters in North American collections but none to George Gordon. In Britain there is little unpublished material that touches on Huxley's work on the Elgin reptiles directly. Because of the thoroughness of the search, it can be said that the likelihood of other parts of the correspondence turning up is slight, unless there are letters in private collections that have not yet been identified.

THE PHYSICAL DESCRIPTION OF THE LETTERS

The Huxley letters are invariably written straight out, without correction or deletion, usually on folded letter-writing paper, often with stamped letterhead (for example, The Geological Survey of England and Wales) and always dated, though the dates are often difficult to decipher. The autograph signature has been cut from one or two of them. No envelopes have survived.

The Gordon drafts are written on small scraps of paper, mostly cut from larger sheets. As drafts they carry many corrections and deletions, particularly because, when Gordon had to choose between frankness and politeness, he always veered towards the latter. The corrections and deletions also show that Gordon sometimes, on reflection, decided not to tell Huxley about fossil finds that fell outside the known scope of Huxley's enquiries.

In many cases, these draft letters of Gordon's are in an extremely fragile state.

EDITORIAL METHOD

1. All the letter texts are reproduced verbatim and in their entirety.
2. Square brackets are used in the texts to indicate words or phrases that have proved indecipherable. It seemed better to indicate an area of transcription difficulty rather than conceal the difficulty by guesswork. These, however, are minor matters since there is no case in which the sense is in doubt.
3. The punctuation has not been standardised, though a dash – the consequence of the type of pen being used – has occasionally been

rendered as a comma or a full-stop where a twentieth-century reader would expect one or the other.

4. Square brackets are also used to indicate words which the sense obviously requires but which the writer omitted.
5. At the head of each letter is given the library reference number. In the case of the Elgin letters, the reference numbers are sometimes wrong, for reasons indicated in the notes, but to permit easy access to the collection these reference numbers have nonetheless been retained.
6. The dates given at the head of each letter are taken directly from the autograph manuscript, except in cases noted. Alterations of date are sometimes necessary in order to recover the correct sequence, either on the grounds of internal evidence, or to amend the transcription errors of the first cataloguers, whose task, it has to be said, given the huge bulk of the Gordon collection, was formidable.

LIST OF LETTERS

In the list which follows the date given is the date the correspondent himself wrote on the letter. In cases where it was difficult to read the handwriting of the correspondent, the date is discussed in the notes to the letter itself. The figures in the right hand column are the reference numbers of the Elgin Museum. On the Elgin letters themselves the reference numbers have the form 'DDW/71/859/33' or 'DDW 71/875/34' and so on. As mentioned above, the first cataloguers made a few errors, so that the correct sequence is not necessarily indicated by the reference numbers. Most letters have two sets of reference numbers but only the more recent are given here.

The abbreviation 'APS' (see letter of 12 November 1866) stands for the American Philosophical Society in Philadelphia.

1858		
4 January	to Gordon	18
14 October	to Huxley	26
15 November	Huxley to Duff	33
16 November	to Gordon	35
18 November	to Gordon	36
19 November	to Huxley	34
20 November	to Huxley	36
9 December	to Gordon	45
28 December	to Huxley	49
30 December	to Gordon	51
1859		
23 April	to Gordon	23

17 May	to Huxley	25
19 May	to Gordon	26
20 May	to Huxley	27
5 June	to Huxley	857/3
9 June	to Gordon	29
9 August	to Gordon	33
15 August	to Huxley	34
21 October	to Huxley	39
29 October	to Gordon	40
November	to Huxley	51
11 November	to Gordon	43
19 November	to Huxley	44
21 November	to Gordon	45
23 November	to Huxley	45
26 November	to Gordon	49
28 November	to Huxley	50
25 November	to Huxley	48
9 December	to Gordon	53
14 December	to Huxley	54
1862		
2 August	to Gordon	16
4 August	to Huxley	18
7 August	to Gordon	19
1863		
22 April	to Gordon	40
24 April	to Huxley	44
1866		
10 August	to Huxley	45
18 August	to Gordon	869/19
October	to Huxley	41
27 October	to Gordon	869/23
12 November	to Gordon	APS
22 November	to Gordon	45
1867		
16 August	to Huxley	16
30 September	to Gordon	20
1 November	to Gordon	23
2 November	to Huxley	24
27 November	to Gordon	28

1869		
7 January	to Gordon	17
11 January	to Huxley	2
17 November	to Huxley	27
28 November	to Gordon	31
1870		
31 March	to Huxley	2
14 April	to Gordon	3
1871		
14 July	to Huxley	17
22 July	to Gordon	18
28 November	to Gordon	32
29 November	Gordon to James	32
1874		
5 July	to Huxley	2
25 July	to Gordon	876/3
August	to Huxley	11
8 August	to Gordon	10
1875		
23 September	to Gordon	34
1878		
22 March	to Huxley	4

To: Gordon
From: Jermyn Street
Dated: 4 January 1858
DDW 71/858/18

I have just been examining your last consignment[1] and I must thank you for the careful way in which the fragments have been packed & lettered.

I cannot satisfy myself however, that there are any teeth among these remains. The [ ] of all appears to me that of bone and those bits that look most like teeth are I suspect fragments of ribs or perhaps of phalange.

I have begun to despair of doing much good with these Lossiemouth fossils.[2] The bone is so tender & fragile that it will hardly retain its form when exposed and the matrix is almost as rotten so that one gets no good natural cast.

Near Findrassie on the other hand, although it is true we get no bone, the rock is so hard and the casts so beautifully sharp that the true form of every part which has left its impression may easily be restored in plaster or grotta pescha.

For these reasons I strongly commend Findrassie as your great hunting ground though of course Lossiemouth is by no means to be neglected.

With one exception all the most important remains of *Stagonolepis* now in my hands are from Findrassie.[3]

Accept my best wishes for the New Year.

Notes

1. As explained above, George Gordon's first packages were addressed to Sir Roderick Impey Murchison, Huxley's superior, whom Gordon had known for many years. Murchison, however, thirty three years his senior, allocated palaeontological tasks to Huxley, being himself busy with other matters.
2. For Lossiemouth and other places mentioned in the correspondence see the Plate 1.

   Throughout the nineteenth century Lossiemouth was an important, bustling and growing fishing port. As it grew, its quarries were absorbed within the town itself and thus cannot be inspected today.
3. According to A. D. Walker, these eventually included the scapula, coracoid, ilium, femur, fibula and dorsal suite.

To: Huxley [but also part of a separate letter to Murchison]
From: Birnie
Dated: 14 October 1858
DDW 71/858/26

Inter alia 'The presumed vertebral column turning out to be a *Cephalapod* gives our locality an additional interest. In reading your printed notice[1], it occurs to me to state that as yet no footprints have been met with at the Covesea quarry.[2] They have appeared only in that of Cummingston which belongs to our county member, and not to Mr. A. Young who is the younger brother of the laird of Burghead'.

Notes

1. Of Huxley's presentation to the Geological Society. It is unlikely that Huxley sent him a copy at this stage in their relationship. No doubt Gordon already had access to the publications of the Society.
2. Covesea is pronounced 'Cowsea'. This is one of the many instances of Huxley working without the benefit of an adequate map, and being nudged away from topographical error by Gordon.

To: Mr. Duff[1]
From: Huxley [copied by Gordon]
Dated: 15 November 1858
DDW 71/858/33

I write a hasty note to say that the packages containing the *Telerpeton* have arrived with everything safe and in good order but we have seen nothing of the slabs with footprints which we understood were on their way some time ago. The matter of the vertebra has cleared itself up, as the supposed [ ] turn out to be the cast of a well marked series of vertebrae agreeing in form with the [ ], and with those which have just arrived. I have but just glanced at the contents of the last package but see they promise well.

Note

1. This is the Patrick Duff referred to in Part One, still active though now sixty seven years old.

To: Gordon
From: Jermyn Street
Dated: 16 November 1858
DDW 71/858/35

I wrote a hurried note to Mr. Duff yesterday acknowledging the arrival of your last valuable shipment but as I see from your note to Sir R. Murchison[1] which he put into my hands some days ago that you particularly request to hear of them. I beg to inform you that they have reached us quite safely and that from the hurried glance which I have given them they seem to contain many valuable pieces of evidence. The cast of the jaw and the vertebrae are most especially important. I have now good evidence of the dorsal and caudal vertebrae of this wonderful beast, but I am greedy of facts and always doubtful of my conclusions until they are backed by overwhelming evidence, so that I hope you will send everything you can lay your hands on. The jaw with teeth[2] is a wonderful thing, but very tantalising for as it is only a cast I cannot tell what was the skeleton of the teeth – (a great point in my enquiry) nor can I be sure of their mode implantation which is no less important.

Can you tell me if any other vertebrate remains have been found in the quarries whence the jaw comes save those of *Stagonolepis*? Has such a thing as a scale of a *Holoptychius* ever been seen there?[3]

The slabs from Mason's Haugh quarry near Cummingston (*sic.*) have not arrived. I trust no mischance has befallen them.

The vertebra question has now been solved. There is no doubt the [ ] vertebra referred to belongs to *Stagonolepis*.

In offering you my hearty thanks for your past exertions let me hope I shall have to thank you still more for one of my great desiderata – a bit of a back of a skull or a fragment of tooth of *Stagonolepis*.

Notes

1. Because George Gordon's correspondence with Roderick Impey Murchison is substantial, it is being prepared for a separate volume. In the division of labour between these two colleagues, Huxley concerned himself *only* with the interpretation of fossils. In his letter to Murchison dated 9 November (DDW 71/858/33) Gordon said: 'I trust you or Prof. Huxley will kindly drop me a note of their arrival'. Perhaps this communication lead to a clarification of who should in future write about what.
2. Probably EM 38 K. See Plate 5.
3. Huxley obviously expected a negative answer to this question, which was designed to establish that the fish and reptilian fossils came from different strata.

To: Gordon
From: Jermyn Street
Dated: 18 November 1858
DDW 71/858/36

I am happy to acknowledge the receipt of the six boxes of whose coming you so long ago advised us and which we began to fear must have gone to the bottom in the late gales.

They arrived yesterday afternoon. The cases were unpacked this morning and by the help of your marks we succeeded in putting together the great slabs. You have done so much for us that it is a shame to criticize any of your arrangements but may I suggest that letters & numbers would facilitate putting fragments of slabs together much better than any other marks?

The footprints are magnificent,[1] and perfectly agree with the conclusions as to the size of the animal which I had drawn from the vertebrae.

Note

1. Huxley's paper 'On the *Stagonolepis Robertsoni (Agassiz)* of the Elgin Sandstones; and on the recently discovered Footmarks in the Sandstones of Cummingstone' was published in 1859, having been previously delivered as a talk to the Geological Society.

   In the published version he said: 'I have intentionally deferred to the end of my communication the description of the footmarks which have been observed upon the Sandstones of Cummingstone, because, although they are certainly the tracks of Reptiles, there is, strictly speaking, no proof whatsoever that they were produced by the particular Reptile, *Stagonolepis*'. This is characteristically tentative. For one thing, the examination of evidence is still in progress. Of the faint indentation parallel to the direction of the footmarks, he says: 'The last mentioned features are only to be made out be examining the tracks very carefully by artificial light'. In a note he says: 'I had not done this when my account was given to the Society'.

   A little later in the paper is the characteristic Huxley remark: 'As to the question whether these tracks were or were not produced by *Stagonolepis*, I will only say that I see no reason for assuming that they were not, while there is some ground for believing that they were so produced'. Because Huxley had hurried to draw a publication from the material Gordon had placed at his disposal, the interpretation could not yet be authoritative. This turns out to have been Huxley's way. He did not wait for the research to be complete before publishing, but rather used the publication as a step towards the understanding of a subject.

To: Huxley
From: Birnie
Dated: 19 November 1858
DDW 71/858/34

I was yesterday favoured with your gratifying intelligence[1] that the box had reached Jermyn Street and that its contents were likely to prove highly satisfactory. On one of the larger casts I think I see casts of teeth as well as that of a large bone with some prints of this bony matter fossilized. Upon your breaking them up I trust the interior of some of the slabs will discover organic casts, altho' they mostly appear on the surface. I do hope that by this time the six cases of footprints from Mason's Haugh quarry, near Cummingston, have got to their journey's end. They left Burghead eight days ago per steamer to Leith, where they were to have been put aboard the London steamer on Saturday. I have been given to understand that the detention arose one week from the boat being loaded with luggage at Dunrobin Castle, and the next from stress of weather preventing the steamer from touching our coast. The assurance that they have been seen by you and Sir R. M. will stimulate me in exploring for a 'bit of the skull or a fragment of the tooth of the wonderful beast' at Findrassie. As I have scarcely expended a third of the sum Sir R. M. authorized,[2] I shall go some day with Mr. Coleman and endeavour to find out the *bed* whence the fossils have been taken. The debris (in which they are now found) has been exposed to the action of the atmosphere for 40 years. During this time true teeth & plates wasted away. No other organism has been met with at Findrassie quarry than those which you have seen, – assuredly no scale of *Holoptychius* has been got there. I hear Mr. Blackie[3] has been to Mason's quarry for some footprints. The quarrymen both at Lossiemouth and Spynie are on the alert to lay hold of anything like a bone or tooth that may [ ] for me.

Notes

1. This presumably refers to Huxley's letter of 16 November.
2. This private arrangement between Gordon and Murchison was independent of the later Royal Society grant.
3. At this stage, Huxley, Gordon and their colleagues were on the alert for rivals. A Mr. Beckles, for example, who went to Mason's Haugh, was dismissed by Murchison as an 'accidental interloper'.

To: Huxley
From: Birnie
Dated: 20 November 1858[1]
DDW 71/858/36

I require to learn that the slabs with footprints have reached you before the 1 of December by which day I trust you will be able to place most of them in their relative positions as they lay in the quarry. I do feel that I did not number the parts so as to enable you to do this easily; but we had had a long day's labour and night was coming on. You will observe that a footprint affects several layers of the bed and puts on a different appearance if distant from the surface on which it is impressed.

I went to Findrassie yesterday to reconnoitre for the best place for the tackman to dig for the fossil bed.

Note

1. The letters dated 19 November 1858 (DDW 71/858/34), 15 November 1858 (DDW 71/858/33) and 20 November 1858 (DDW 71/858/36) are all written on a single now battered leaf of paper. The last, partly erased, partly illegible section of the letter dated 20 November 1858 has been separately marked as DDW 71/858/33, though this number had already been used for Huxley's letter dated 15 November (above).

   Although the letter is dated 20 November, it was probably replaced by what is now the preceding one. If letter 33 was written on 20 November, then letter number 34 cannot have been written on 19 November. Gordon must have dated one of these drafts inaccurately and in all likelihood we have here an example of a letter that was not sent, although it expresses his impatience. He had sent such a treasure to London. What had happened to it?

To: Gordon
From: Jermyn Street
Dated: 9 December 1858
DDW 71/858/45[1]

Thanks for your welcome announcement of more bones from Lossiemouth. It is from these that I hope for most.[1]

The evidence comes out and changes & changes in favour of the Crocodilian nature of *Stagonolepis* (Agassiz spelt the word this way though I suspect it is wrong & should be *Stegonolepis*) and I only cry now 'A tooth! A tooth! my kingdom for a tooth!' If I could only make sure that this cast of a jaw you just sent me belonged to our reptile it would be something, but as you say in the North there is something 'uncanny' in the look of it.

Accept my best apologies, my dear sir, for having unconsciously unfrocked you. I ought to have remembered what 'Manse' means.

Notes
1. This is in Huxley's hand, though the signature has been cut from the letter.
2. He has forgotten what he said on 4 January earlier in 1858.

To: Huxley
From: Birnie
Dated: 28 December 1858
DDW 71/858/49

Yesterday I gave to the Railway Carrier a small box addressed to Sir R. Murchison containing four bits of the Lossiemouth rock with remains of *Stagonolepis* in each.

The parcels marked A. & B. contain forms the likest to teeth that I have as yet seen. I am most anxious to hear your opinions of them, and am sure you will early favour me with your verdict. I have marked the fragments so that I hope you will be able to replace them. The two bits themselves were contiguous; but the quarryman, before he gave them to me and when my back was about, unfortunately broke them that I might see the contents thus better. One '*tooth*', shall I say, in parcel A. would according to my recollection exactly fill up one of the 'uncanny' looking casts you got some time ago. Some of the forms in parcel B. look more like teeth and I see some fluting on them.[1]

This box which should reach London about Saturday will [be] the last contribution it will be in my power to send you for some time, as the Elgin Literary Society wish to have something to exhibit in the Museum to the multitude of *Stagonolepis* hunters which your paper of the 15th will send down on us next season.[2] However the directors I am sure will at once send you for inspection anything that may now come into their hands that is new or remarkable so as to enable you fully to reconstruct the animal.

P.S. Did you get the last half of the despatch of which I some time ago gave you notice?[3]

Notes

1. The field worker triumphant. You want stone from Lossiemouth after all? Here is some. You want a tooth? Here's a tooth. Huxley had said in his previous letter that he was in the midst of writing his paper, so in all probability it was this consignment that put him in a position to write the 'Jaw and Teeth' section of it. On *MII* 112 he concludes: 'If this jaw belonged to the same animal as that to which even the largest of the vertebrae belonged, the size of the teeth is remarkable, for the longest is considerably more than twice as long as the centrum of such a vertebra'. Though there is no proof that the jaw and teeth are those of *Stagonolepis*, he believes they are. As usual he is trusting techniques of comparative anatomy, not having seen the specimens *in situ*. He also believes that *Stagonolepis* is Crocodilian. On the one hand he says that he is at a loss to find an 'exact parallel for this peculiar combination of characters in any group of recent or fossil *reptilia'* (the Elgin *Stagonolepis* was after all unique at the time of writing); on the other hand, the data does not enable us to deduce 'the precise age during which the Elgin fossil lived'. It is interesting to read the section of the paper entitled 'Affinities of the Stagonolepis' in conjunction with this.
2. Whether the museum was swamped with visitors in 1859 is uncertain, but certainly what seemed Huxley's outrageous suggestion that there had once been crocodiles in the north of Scotland attracted a lot of attention. George Borrow, for example, has the narrator of *Wild Wales* rest by a lake in the hills west of Festiniog. 'There lay the lake in the low bottom, surrounded by heathery hillocks; there it lay quite still, the hot sun reflected on its surface, which shone like a polished blue shield. Near the shore it was shallow, at least near that shore on which I lay. But further on, my eye, practised in deciding upon the depths of waters, saw reason to suppose its depth was very great. As I gazed upon it, my mind engaged in strange musings. I thought of the afanc, a creature which some have supposed to be the harmless and industrious beaver, others the frightful and destructive crocodile'. As to the word 'afanc', Borrow says, is it not 'a fitting brother of the Arabic timsah, denoting the dread horny lizard of the waters?' and later in the same paragraph 'If crocodiles – ever existed in Britain, and who shall say that they have not, seeing that their remains have been discovered, why should they not have haunted this pool?'. Of course there were a lot of reasons for crocodiles not having haunted a mountain pool in igneous

rock. Nonetheless, Borrow's literary imagination, working in a different way from Huxley's, had seized on the essence of a discovery, though without the desire for the type of proof Huxley required.
3. May this have been the re-written draft of 20 November 1858?

To: Gordon
From: Jermyn Street
Dated: 30 December 1858
DDW 71/858/51

I fully meant to have acknowledged the receipt of your last package but I fear that as I was in the midst of preparing for my paper at the Geological Society it must have slipped out of my mind. I have not as yet, found anything new in it but many of the blocks require to be broken up.

I look forward with anxiety to the new arrival & will let you know at once when it reaches me.

You are quite right about the derivation & spelling of *Stagonolepis*.[1] I had fancied it meant 'covered scale' from the overlapping edges of the scutes, but I set myself right before the meeting of the Society.[2]

The Directors of the Elgin Museum have been exceedingly kind & liberal to us, that I should be the last man in the world to complain of their keeping anything they please, but, in the interests of science, I hope that if anything like a bit of skull or teeth, or bone of the feet or pelvis, turn up, they will let me see or know of them.

You will be surprised at the amount of evidence that has already turned up and, even now, I have in my hands the material for a fine monograph – which if nothing unforeseen occurs will I trust be out by this time next year. By that time I hope that the age of the Beds will be settled past controversy.

Notes
1. A clear example of a letter from Gordon not having survived.
2. When he delivered the paper entitled 'On the *Stagonolepis Robertsoni (Agassiz)* of the Elgin Sandstones: and on the recently discovered footmarks in the Sandstones of Cummingstone'. *Quarterly Journal of the Geological Society* (1859), 440–60; *MII* 94–119.

To: Gordon
From: Jermyn Street
Dated: 23 April 1859
DDW 71/859/23

I write a hasty note (to save[1] the post) to tell you that your last shipment has arrived quite safely.

I have put the fragments together & the hasty inspection which I have made of them is sufficient to enable me to say that the creature *is not Stagonolepis* & *is* one of the most remarkable anomalous Reptiles I have seen.[2] Pray continue your search. The present specimen resolves many doubts I had about the first & every fragment is of value.

I shall be able to say nothing decided about the creature for some weeks. There is nothing for it but to sit down knife & chisel in hand, before the fossil for a week or two & fairly compel it to reveal its secrets by dint of scraping.

Notes

1. The *O.E.D* gives 'to be in time for, manage to catch' (1732) and quotes Canning: 'I have but a moment to save the post'.
2. This is the first mention in the correspondence of *Hyperodapedon*.

To: Huxley
From: Birnie[1]
Dated: 17 May 1859
DDW 71/859/25

They (the specimens) belong to the Elgin Museum, whose Directors, anxious that the geological rarities of our country should be *known* and seen at *Aberdensis* [ ] fondly hope that you will favour the Geological Section of the British Association with some account of them. But be this as it may, the Directors expect that, after you have made them by chisel to reveal their contents, the specimens will be exhibited in the Granite City; as they will be among the chief objects of Geological interest, collected in the north for the forthcoming meeting there.[1]

Notes

1. Gordon says in a note that he has copied only what he calls the 'principal part' of this letter. Evidently his patience is wearing thin. In his anxiety to publish and to make his mark, Huxley perhaps forgot that a genuinely disinterested person was doing some of the work for him. His position was similar to that described by Donald Johanson in his book *Lucy's Child*, where there is discussion of the ownership of fossil remains found by foreign researchers in East Africa. Johanson has Tim White agree with the idea that fossils belong to the country in which they are found: 'Someone gets a permit, excavates, finds a fossil. Then off it goes on loan, to be studied in some other country, written up in a foreign journal by authors with foreign names . . . you can't blame the people here for wanting to take control of them.' No more could you blame Gordon for wanting the fossils back in Scotland. See also the letters date 10 August 1866 and 11 January 1869.
2. In 'The Geology of the Lower or Northern Part of the Province of Moray', Gordon had already given an indication of his own position. He speaks of

the 'hiatus' between the ridge between Bishopsmill and Alves. He knows that fish and reptilian fossils are not found in the same stratum. He is familiar with the fossil *Stagonolepis Robertsoni* found near Findrassie by his friend, Alexander Young. And he believes that it belongs to a 'higher class of reptiles'. It was not Huxley alone who was thinking about the implication of the fossil finds. Interestingly Gordon says, *pace* Huxley: 'We trust that at least a cast of this slab, and that other fossils from this now so interesting locality, will be sent to Aberdeen, by the time that the British Association meets in that city [Granite City], for the inspection of those able to decide as to the character and position which these animals should have assigned to them among the vertebrae'.

To: Gordon
From: Jermyn Street (on Geological Society paper)
Dated: 19 May 1859
DDW 71/859/26

Thanks for your excellent news. I propose to go to the Aberdeen meeting[1] for the precise purpose of giving a 'screed' about your famous Reptiles whose interest '[ ].[2] Of course the specimens shall come and I must beg you to thank the directors for their consideration in placing the resources of the Museum at my disposal, especially considering the claims which have (as I hear) been put forth in favour of the 'Great National Collection'.

Your last package turns out to be one of the most interesting which has yet made its appearance. It is a *Lacertian* having some relations with *Placodus*[3] *with Rhynchosaurus*,[4] with (alas for Sir Roderick!) unquestionably *Triassic* forms, and if you will forgive me for taking your name in vain, I call it *Hyperodapedon Gordoni*. The generic title is given on account of the pavement of teeth in its palate.
[after the signature]
I will duly acknowledge the new arrivals.

Notes

1. For the annual meeting of the British Association where he gave a talk called 'On the newly discovered Reptilian Remains from the Neighbourhood of Elgin'.
2. [the Latin tag is as yet indecipherable].
3. In W. E. Swinton's *Fossil Amphibians and Reptiles* (British Museum, 1973), it is said that the Placodonts had a 'terrestrial ancestor of typically reptilian appearance with a characteristically land-reptile skeleton, so that when they went to sea they did so only by re-adapting their structure to the demands of aquatic life and not be recovering the primitive structures that their ancestors once had'.
4. To the extent that the correspondence provides evidence about how Huxley thought and worked, this letter is crucial. Huxley's first instinct is to establish the new fossil's systematic placement. Reasonably enough

perhaps, he relates it to what he already knows, despite the vagueness of 'having relations with'. He says 'unquestionably Triassic' only because he had not read about any pre-Triassic reptiles.

To: Huxley
From: Birnie
Dated: 20 May 1859
DDW 71/859/27

I am indeed gratified by having my name associated with that of the new Lacertian, and beg to thank you warmly for the compliment. Your kindness to me is the more pleasing to me, as the fossil belongs to Moray, my native Province, whose natural history in several of its departments I have for many years been endeavouring to make known. It delighted me to think that the interest of the scientific world in the locality is likely to be increased.

I must say however that it was the quarrymen who were the discoverers of the specimen of the reptile which I sent you – although on hearing of this discovery from Mr. Leith the tackman, I repaired to Lossiemouth and secured it – one portion in the quarry and another at a distance whence it had been carried for housebuilding. They are now so alive to the appearance of any fossil looking matter in the rock that nothing of the sort is likely to escape them. This is so far well; for we must have something more from the formature before we sever it from the Old Red and give it up to the Triass. If we must do this it will be with a pang but *Magna est Ventura* etc.

To: Huxley
From: Birnie
Dated: 5 June 1859
DDW 71/859/3[1]

I have been expecting to hear of the large box, despatched some three weeks ago, having arrived in Jermyn Street. It is too large & heavy to have been accidentally lost & will certainly get to its destination however slowly, if not already there.

But my purpose in writing at present is to account for a paragraph that has unfortunately got into the local paper concerning the *Hyperodapedon*. I showed your letter to some of my co-directors of the Elgin Museum in order that they might receive as if from your own hand your thanks for the use of the specimen. The perusal of it highly gratified them all. One of them, Mr. Martin[2], from the best motives & unknown to me sent anonymously (?) a short but not very accurate notice of the discovery to the two Elgin papers hinting at its Triassic resemblances[3], one of which inserted it and from which it has been quoted into other Scotch newspapers. The editor of the other luckily consulted my friend Dr. Taylor[4] who got them to withdraw it.

If it chance to meet your own or the eye of any of you from Jermyn Street you will remember this explanation of the true account of the mishap.

Notes

1. Often it is difficult to distinguish between 7 and 9 as Gordon wrote these figures. Although this letter is catalogued as 1857 it is virtually certain that it was written in 1859, when the *Hyperodapedon* fossils were discovered.
2. In the essay 'On the Geology of the Lower or Northern Part of the Province of Moray', Gordon says that 'Mr. Martin, of Anderson's Institution, Elgin, in 1835 wrote an essay on the geology of Morayshire, for which he obtained a gold medal from the Highland & Agricultural Society of Scotland'. It was published in Volume V of the Society's *Prize Essays and Transactions*.
3. See the *Elgin & Morayshire Courier*, 8 April 1859, p. 6.
4. This is the Dr. Taylor of the Indian Civil Service to whom Gordon refers in his article in the *Edinburgh New Philosophical Journal*. Taylor subsequently left many of his fossil finds to collections in Aberdeen where Huxley may have seen some of them.

To: Gordon
From: Jermyn Street
Dated: 9 June 1859
DDW 71/859/29

I had imagined I had long since acknowledged the safe arrival of the last cargo – a very valuable & interesting one by the bye – and I must apologize for not having done so.

It would have been as well to have said nothing about *Hyperodapedon* at present, but as the notification is published it cannot be helped.

Happily I can see no reason to alter any opinion I expressed in my note to you.

The only question in my mind now is the possible identity of *Hyperodapedon* with *Rhynchosaurus*.[1] I don't think it at all likely that this will turn out to be the case but I hope for some fresh material on the *Rhynchosaurus* with [which] to settle the question.[2]

Notes

1. Huxley knew Richardson's work on *Rhynchosaurus* which in Britain had first been found in the New Red Sandstone in Shropshire in 1851.
2. Huxley's signature has been cut from this letter.

To: Gordon
From: Aberdour, Fife
Dated: 9 August 1859
DDW 71/859/33

You see I have shifted my quarters a good deal nearer to you though we are still a long way apart. I shall be here with my family until the meeting of the Association. Before leaving London I had a large [ ] of the specimens which you had so kindly lent us from time to time, packed up. Sir Roderick Murchison thought they had better be sent direct to Aberdeen but before giving any order about them I thought it best to ascertain the wishes of the Committee of the Elgin Museum.

Will you therefore be so kind as to let me know, as soon as may be convenient, where the specimens should be sent?

Is it likely that there is anything to be gained by my hunting over your ground before the meeting? If so I would run up to Lossiemouth by the steamer before going to Aberdeen.

To: Huxley
From: Birnie
Dated: 15 August 1859
DDW 71/859/34

Owing to my own and the absence from home of the Directors of the Elgin Museum, I have been unable earlier to reply to your note of the 9th inst.

Your proposal to send the fossils direct to Aberdeen (and to the care of Professor Nicol[1] of Marischal College) meets the entire approbation of the Directors.

Beyond some blocks containing caudal vertebrae & plates of *Stagonolepis*, obtained at the Lossiemouth quarries last week by Dr. Taylor,[2] nothing new has turned up at that or at any other of our reptilian localities.

We are not able to hold out to you the inducement of new matter for your undertaking a voyage from the Firth of Forth to that of Moray.[3]

Still, if your time permitted of such a short recreative trip to the North, the Directors would be grateful to know you had inspected their Museum & examined the original habitat of these singular creatures which you have now made famous.

I should be delighted to go down & meet you in Lossiemouth and give you any assistance in my power.

Have the goodness to say if we may encourage the hope of your coming, as in that case the transference of the Museum fossils (from

Elgin to Aberdeen) would be delayed until you had seen them at Elgin.

Notes

1. James Nicol, Professor of Geology at the University of Aberdeen, was a person with whom Gordon frequently corresponded. His *Guide to the Geology of Scotland* had been published in 1844.
2. It is difficult to determine how well Huxley knew Taylor, and Taylor's collection. A number of letters in the Elgin collection to Gordon from G. E. Roberts refer to Taylor and, in 1863, to Huxley having seen Taylor's fossils. But which specimens he saw is unknown.
3. There is a letter in the Elgin Museum from Gordon to J. H. Hulke (DDW 71/844/7) which presumably should be dated to the 1850s, not 1844, since it refers to his having sent scutes and bones to Huxley 'which he described'. He says there is nothing like a pelvis in the Museum (which evidently Hulke had asked for) but that 'there is a good specimen of a reptile got a year or so ago'. This letter is being referred to now in the present note to DDW 71/859/34 because it seems likely that in 1859 Gordon was not as familiar with Huxley's work habits as he later became. Huxley was never in London during the latter part of August. The 'good specimen of reptile' *may* have been of the *Hyperodapedon*; equally it may have been a specimen which Huxley never saw. Did Gordon refrain from offering Huxley the inducement to visit Moray because of his dealings with other people who were also interested in the region?

To: Huxley
From: Birnie
Dated: 21 October 1859
DDW 71/859/39

Robert M'Andrew Esq., of Isleworth, in the course of ten days or so will show you a small bit of *Stagonolepis*, I think a portion of both jaws with teeth. We got it in the Lossiemouth quarry where some more pieces have turned up, but had been a good deal broken up before I saw them. I have got what remained taken up to Elgin but I question whether there will be anything worth the sending you.

Mr. Patrick Duff at the same spot of the quarry and if not of, assuredly among the remains of *Stagonolepis* got a portion of a tooth which he showed me and which he intended to send to London by Mr. M'Andrew for your & the inspection of *other palaeontologists*.[1] Mr. Duff's small bit of tooth was compressed and & its edges *were serrated*. I did not observe the structure in any of the other teeth. Does this not shew an affinity to the Thecodonts[2] – or did both forms of tooth occur in the same mouth?

Mr. M'Andrew will give you a bit of sandstone to which you will kindly look when your other more important avocations permit. I suspect that the tabular harder parts are more concretions. It is from the sandstone near Forres.

Notes

1. The *other palaeontologists* included Richard Owen, co-founder of the Palaeontographical Society and already an established figure in the field Huxley was entering. Owen's *A History of British Fossil Reptiles* was mentioned in Part One, but he had also published books like *On the Archetype and Homologies of the Vertebrate Skeleton* (1848) and *On the Classification of Geographical Distribution of the Mammalia* (1859) which, with other publicatons, were very close indeed to Huxley's interests. The two men were rivals. A book like Owen's *Paleaontology* (Edinburgh: Adam and Charles Black, 1860) quickly gives the twentieth century reader an inkling of why Huxley could not abide Owen, since Owen is unwilling to see palaeontology as a reason for not believing in Christianity. On p. 3 he says that 'perhaps the most important and significant result of palaeontological research has been the establishment of the axiom of *the continuous operation of the ordained becoming of living beings*' and he ends the book with the sentence 'Thus the highest generalizations in the science of organic bodies, like the Newtonian laws of universal matter, lead to the unequivocal conviction of a great First Cause, which is certainly not mechanical'. Neither statement would have been acceptable to Huxley.
2. Thecodont: 'an extinct family of saurians having the teeth fixed in sockets in the jaw-bone'. The first usage of this term recorded in the *OED* was Owen's in 1842.

To: Gordon
From: Jermyn Street
Dated: 29 October 1859
DDW 71/859/40

Mr. MacAndrew[1] (an old and valued friend of mine, by the way) called on me yesterday and left with me the specimens to which you refer. The bit containing part of both the upper & under jaws is immensely interesting and I cannot thank you enough for it. Mr. MacAndrew mentioned to me the other tooth (from Mr. Duff) stating that in accordance with Mr. Duff's wish he had left it at the British Museum to be shown to Prof. Owen & afterwards sent on to me.

For very good reasons which you will readily understand without my troubling you with them at length, I immediately declined having anything to do with the specimens and I wrote to Mr. Woodward in whose charge it is, requesting him to return it directly to Mr. Duff. I also wrote to the latter gentleman to tell him what I had done.

I should be very sorry to offend Mr. Duff and I have no doubt his intentions were excellent but his method of procedure, is to say the least unusual.

I do not know to what 'Paper' of Owen's you refer but if you will kindly let me have the reference to it I will look into the matter. To my knowledge he has never seen any tooth of *Stagonolepis*[2] save those that were on the table at Aberdeen.

I cannot make out anything organic in the curious body from Forres.

Is it true that you still hold to the Devonian nature of the Reptiliferous beds?[3]

Notes

1. Robert MacAndrew wrote to George Gordon on the same day. 'I yesterday gave the Lossiemouth fossils to Huxley who is much pleased with that containing teeth of both upper and lower jaw. Our friend Mr. P. Duff made a mistake in wishing his specimen to be shown both to Huxley and Owen. After calling at the Museum twice without being able to meet the latter, I left the fossil in charge of S. P. Woodward with instructions to show it to Owen and afterwards forward it to Huxley, but he (Huxley) hearing it was in Owen's hands would have nothing to do with it – I of course do not mix myself up in such squabbles.'
2. It is difficult to determine whether this is the same specimen as E. T. Newton referred to years later. In a letter dated 1 June 1893 (DDW 71/893/14) he told Gordon: 'You will be glad to know that the specimen that was with Woodward, has been handed to me, and has turned out even better than was expected; the only parts missing are the neck and the forelimbs'. Did Huxley miss an opportunity here as a result of his animosity towards Owen?
3. Huxley's letter to Sir Charles Lyell dated 10 August 1859 [Imperial College] is of importance here. Evidently he had written to Lyell after the British Association meeting to have his opinion on the geology of Moray. He finds Lyell's letter 'highly satisfactory' and explains: 'I did not like to be dogmatic upon the point, the value of the reptilian evidence not being so certainly defined as that derived from moluscs, but I must say I would have been greatly astonished if after all the Elgin beds had turned out to be Devonian'. Perhaps the reason for Huxley's disingenuous letter to Gordon is that each man by this time knew the other's position, especially because they had met at the Aberdeen meetings.

To: Huxley
From: Birnie
Dated: November 1859
DDW 71/859/51

I am glad you were so pleased with the bit of double jaw which Mr. M'Andrew gave you. I hope it will show you (if such exist in the Stagonolepis) serrated edges in the teeth.

I do not exactly remember what I said in my last note regarding serrated teeth; but my meaning was that as Professor Owen in his paper, read at Aberdeen and since printed in the *Athenaeum* of 1 October last, mentions this form of tooth only when he describes the Thecodonts, all of which he calls Triassic, you, Palaeontologists, would perhaps find in it another argument for holding our Lossiemouth fossils to be Triassic.

The portion of tooth with serrated edges (?) Mr. Duff found *among* the remains of *Stagonolepis*: and it most probably belongs to that animal, – if reptilian teeth differ so widely in the same mouth as to present some with & others without serrated edges. Or, if no such wide difference is known to exist, this tooth may indicate the existence of another animal.

You ask me if it be true that I still hold by the Devonian nature of the reptiliferous beds. It is so, – at least to this extent, that the stratigraphical and lithological evidence so strongly supports that view, that I do not think it yet undermined by the strongly Triassic-looking reptiles.

In short I hold the age of the beds to be an open question, on which I look for more evidence. Of such additional evidence, when it does appear, if in my power, you shall have an early notice and inspection.

Much stress is laid by some on the Linksfield beds being, as Mr. C . . . probably correctly calls them 'Lias running into Trias'; but you must see that this foundation is not strong when Sir. C. Lyell thus describes a specimen taken from the underlying Cornstone there. I quote his own words as he kindly has written them for a specimen in the Elgin Museum. 'Fundamental limestone of the Linksfield quarry, with glacial strains (?) running W by N and E by S, covered by glacial boulders, clay and this by Lias'.[1]

Living remote from cities, Libraries, Periodicals etc. I am not so fully 'read up' in the history of Geology as I should desire, but am I far away when I think that this history tells us that very few vestiges, if any, of true Reptiles appear for the first time prior to the Triassic age? Then, if I am correct in this, the known Triassic reptiles would be those which the Elgin reptiles would be expected most to resemble if they really be Devonian; as those Triassic forms would be nearest to them in the geological scale.

And suppose (for the thing is not impossible) that you heard from unquestionable authority that a veritable reptile had turned up in the coal of Newcastle or in the Old Red of Devonshire would you not at once suspect that it would more resemble the *Rhynchosaurus* or the *Labyrinthodon* than any other because they were newest to it in time?

Some say our beds are newer than the Trias; and the known Triassic reptiles in other localities (being the oldest) come up prominently and by themselves as a standard of comparison for the Elgin reptiles, – whatever their geological age may in fact be.

What I now state, I allow, is not evidence in either direction; but it may prevent us concluding that a *reptile* must be Triassic because it resembles Triassic more than other forms, with the same confidence as we must conclude that *a fish* is Triassic because it resembles Triassic more than other forms. For in the latter case, that of the fish, we have elements for contrast & comparison from every formation, whereas in the former the paleozoic rocks give us no rocks wherewith to compare the discovery.

Had reptiles of various forms been previously found scattered through the Silurian, Devonian, Carboniferous & Permian Strata, if an affinity had been found in the Elgin reptiles to the Triassic more than to other forms than to these supposed paleozoic reptiles, the palaeontological argument would have been *much much* more strong as to their age.

Forgive me for withholding my assent to the 'dictum' 'they must be Triassic' until more light be got on this, to me at least, *a questio vexatio*. I trust also you will forgive this rather lengthy note the reading of which, even if I have made myself intelligible, must encroach on your valuable time.

P.S. There is abundance of the concretionary sandstone from Forres – so break up the best sent you to its very core.

Notes

1. Lyell had visited Elgin after the meetings in Aberdeen.

To: Gordon
From: Jermyn Street
Dated: 11 November 1859
DDW 71/859/43

Many thanks for your long and valuable letter. The peculiar character of the *Thecodonts* referred to by Prof. Owen was long ago discovered & described by Riley & Hatchbury, but he would be a bold palaeontologist who would build much on such a character as the serration of the teeth, a character which is by no means peculiar to the *Thecodonts*. The teeth in the specimen you sent me are exactly like those I have already described in *Stagonolepis* but the implantation in distinct aveoli (?) is better there.

I have very great doubts whether the large jaw with teeth from the Findrassie beds belongs to the same Reptile & I wish those Findrassie beds could be more thoroughly worked out.

I have broken up the specimen from Findhorn but am not much the wiser. Two or three faint brownish lines (which might or might not indicate organic matter) is all the portion so far revealed to me.

I see you lay no great weight on Sir C. Lyell's hypothesis that the Linksfield Trias has undergone a landslip over a glacial drift.[1]

Your palaeontological argument in favour of the Devonian nature of the Elgin Reptiles appears to me to be open to objection to a certain extent. The only true reptiles (excluding Batrachianas) older than the Trias, at present known, are the Permian *Protorosauri* whose type is the famous 'Muringian Monster'. Then comes the whole range of Carboniferous rocks in which as yet no reptiles higher than *Batrachia* have been found.

Then come our (supposed) Upper Devonian Reptiliferous Beds, where you find a Saurian Reptile (*Hyperodapedon*) which is not like the Muringian Monster and a Crocodilian (*Stagonolepis*) which is not like any known Permian or Triassic Reptile but resembles with wonderful closeness the *Rhynchosaurus* of the Lower Lias. Thus you see it is not quite true that the Triassic reptiles are the nearest in time to the (supposed) Devonian forms.

I do not mean to say that what we know of Palaeontological Law at present warrants me in saying that the Elgin Reptiles *must* be Triassic – for neither of them are of known Liassic or Triassic Genera.

Nor would I pretend for one moment that they *cannot* be Devonian – I, for one, see no *cannot* in the case. And if I heard of the discovery of such reptiles in Siluria or Longnesel (?) itself I should have no *a priori* reason to urge against the fact.

But the case seems to me to stand thus. It is granted that the stratigraphical evidence is not decisive either way. The Palaeontologist must therefore disregard it & be guided solely by the affinities of the fossils. Now the nearest allies of *Hyperodapedon* & *Stagonolepis* are admitted on all hands to be the Liassic or Triassic *Reptilia*.

Therefore until the contrary is proved, the presumption is that the Reptiliferous beds are Triassic at latest.

I am quite ready to veer round to any extent when fresh evidence is brought forward, but at present I do not see my way to any other conclusion than that which I have laid before you.

I have been interrupted a half a dozen times by people coming on very different business, while writing this letter but I hope I have made my view of the case intelligible. Any way I hope you will see that I hold my opinion lightly in hand, as one earnest for truth only should do.

I ride with neither the Capulets nor the Montagues, my Juliet being *Fact*.

Notes

1. The effects of glaciation, though understood by Lyell, were still being disputed by others. Murchison was among those who refused to accept the evidence of the ice-age.

To: Huxley
From: Birnie
Dated: 19 November 1859
DDW 71/859/44

Many thanks for your kind & interesting letter, which, without destroying my tendencies to the Old Red, gave me much instruction.

I now write to ask you to break off, toward the point, a small portion of the *outer layer* of the enamel (dentine matter) in one of the teeth which

Mr. M'Andrew gave you, and look narrowly at the exposed part for serratures on its edges.

Among the parcel of broken and fragile pieces which I got from the very same spot of the quarry – all belonging to the same individual most likely, I find a bit of jaw with one tooth at least in it. Trying carefully to clean it off the matrix, a film of the tooth broke off and exposed serratures, very like those I saw with Mr. Duff although not on so large a scale.

If you cannot detect this appearance in your specimen, be so good as to let me know & I shall forward my specimen at the first opportunity. The one I have I am persuaded shows the serratures towards the front – [the incomplete draft breaks off at this point]

To: Gordon
From: Jermyn Street
Dated: 21 November 1859
DDW 71/859/45

I have excavated the matrix about several *Stagonolepis* teeth so carefully & fully that I have no doubt about their character. The crowns are swollen at their base but taper towards their apices. Their transverse section is oval the oval being much more rounded below than towards the apex. As to the apex in fact, the teeth are sharp & sometimes rather raised & one or other of their edges is sometimes a little [ ] – less so, if anything, than my rough figure represents.

There is nothing to be termed a serrature on these edges. No trace of such a thing is visible either in the teeth themselves or in their casts.

A very characteristic feature of the teeth is that the convex surface of the crown is marked by a number of irregular shallow longitudinal grooves, in consequence of whose existence the edges of the transverse section are not even but slightly [ ] curves.

I shall be very much obliged if you will send up any specimens, however fragmentary, of serrated teeth.

To: Huxley
From: Birnie
Dated: 23 November 1859
DDW 71/859/45 [this draft is written on the back page of Huxley's letter of 21 November 1859 and has the same Elgin reference number]

I perfectly agree with all that you say and *figure* regarding the teeth of the *Stagonolepis* in their entire state, – that is when the enamel or outer layer is present; but do oblige me (however sceptical you may feel) by removing a small portion of this enamel or natural outer layer at or near the apex, and then tell me if you see any serratures on the edges of the underlying dentine matter.[1]

Note

1. In an earlier draft of this letter, written on the same sheet of paper, but then heavily deleted, are a further five lines referring to the serratures of the tooth Mr. Duff sent to Professor Owen.

To: Gordon
From: Jermyn Street
Dated: 26 November 1859[1]
DDW 71/859/49

I have removed some of the enamel in a *Stagonolepis* tooth, but I see no serratures or anything more marked than what I mentioned to you in my last note. You recollect of course that the serratures of *Thecodontosaurus* are formed of the enamel. However I should like to see the specimens which puzzle you very much.[1]

Let me offer my congratulations on the well deserved compliment the University has just paid you as I learn from Sir Roderick.[2]

& believe me
my dear 'Doctor'

Notes

1. The date of this letter should probably be 24 November 1859, since the next one is obviously an answer to it.
2. Short of going to see the specimens in Owen's care in the British Museum.
3. Colin McLaren, present special collections Librarian at Aberdeen, informs me that there was no citation on the occasion of George Gordon's award of an honorary degree.

To: Huxley
From: Birnie
Dated: 25 November 1859
DDW 71/859/48

Many thanks for your kind and hearty congratulations on my receiving an honorary degree from my Alma Mater; although I must put in an exception as to the amount of credit which you would give the Senators for this step which they have been pleased to take.

You well know it is an honor (*sic*) I would never in any way solicited or been in any degree anxious for – as in some respects I would have been better without it. But it came upon me in a way I could not well decline.

[at this point a sentence, written between the lines & partly deleted, is illegible]

But however I may estimate it I prize it highly as it affords me an occasion of [having] you and like kind of friends gratifying me in the expression of such sentiments and feelings as I find in your letter just received.

To: Huxley
From: Birnie
Dated: 28 November 1859
DDW 71/859/50

The piece of rock which contains the tooth of *Stagonolepis* to which I have been referring in my last two letters, is too large for the post and too small to be sent by itself per Steam-boat; so I purpose (tomorrow if possible to catch the weekly boat) to despatch it via Aberdeen in a box filled up with a few other things likely of no value but put in for the purpose of giving weight.

Mr. Duff and one or two others accustomed to look at such things, having seen the tooth today, agree with me that there are serratures, or strongly and regularly marked serrosities (?) on the edge laid bare of enamel. But I leave you to form your own idea of it when it reaches you. I put a little gum on the tooth which can easily be washed off. I make it over to you as well as the contents of the box.

There are two or three bits of sandstone with small concretions (one of them full of pyrites) which I think are organic, caused likely by some part of the *Stagonolepis* having once been there. Indeed on one of them I see bits of scale or bone.

I am persuaded that the tooth I send you is exactly (tho perhaps not so large) of the same nature and form as that which Mr. Duff sent to London and which he says is serrated and the tooth of a Thecodont. I believe it (for I saw it) after having examined my own specimen, to be nothing more or less than a small section of the upper half of a tooth of the *Stagonolepis* denuded of its enamel, and not the tooth of a new genus or even species.

To: Gordon
From: Jermyn Street
Dated: 9 December 1859
DDW 71/859/53

Your specimens have reached me quite safely and I can assure you that 'the tooth' is without doubt a very good specimen of that part of a *Stagonolepis*. The 'serrations' are of exactly the same character as those I have seen in other specimens only they are somewhat more regular – & they certainly bear no resemblance to the serratures of the 'Thecodonts', which involve the external enamel.

It is most singular that with all these remains nothing should turn up but *Stagonolepis* & *Hyperodapedon* oh for a shell of a fish! Surely these reptiles were not Psaurnophagi.[1] Brutes such as these must have required a considerable dinner now & then one would think.[2]

Accept my best thanks for thinking of my wants & sending the Cornstone specimens. Both are very good & instructive.

I do not see the Elgin Museum upon them. I shall mark them as yours. Everything unmarked is sure to find its way into our cases – 'the bourne from which no traveller returns'.

I enclose the abstract of a lecture of mine which may perhaps interest you.[3]

Notes

1. That is, eaters of the speckled ferns that abounded in that era.
2. M. J. Benton refers to Walker's earlier idea that, while smaller reptiles may have lived among the aeolian dunes, the *Stagonolepis* and *Hyperodapedon* 'must have sought plants in the adjacent flood planes or around interdune pools' (Benton, p. 690). Late Triassic food chains in the Elgin region cannot be established 'since no plants, invertebrates or fish have been discovered'. Because the *Hyperodapedon* had a 'precision-shear bite', a powerful tongue, a beak, and limbs with claws, Benton believes its diet must have consisted of 'tough, but not hard, plant material'. The skeletal reconstruction suggests that it was not capable of rearing up on its hind legs to reach the leaves of trees. Others have suggested that the diet may have included molluscs.
3. Possibly the abstract of 'On the Dermal Armour of *Crocodilus Hastingsiae*', *Quarterly Journal of the Geological Society*, Volume 15 (1859): 678–80.

To: Huxley
From: Birnie
Dated: 14 December 1859
DDW 71/859/54

I am glad the box reached you; that its contents gave you interest & that you have seen on the 'tooth' all that I expected. I had a great inclination to break up the specimen as I thought I saw an indication of two other teeth in it; but I leave this for you to do, in the hope that you will find still better marked serratures under the enamel.

The question now is were the serratures on the tooth (or rather part of a tooth) sent by Mr. Duff to London, of this same sort? I have a strong opinion that they were so, and not the true serratures of a Thecodont.

The *Coccosteus* (got twenty years ago with Dr. M[1] . . . in Nairnshire) and the other contents of the box *were mine*; they are *now yours*. In two or three amorphous bits of stone, I do hope you will be able to detach at your leisure something organic. They were got among the remains of *Stagonolepis*.

Note

1. No doubt Dr. Malcomson.

To: Gordon
From: Jermyn Street
Dated: 2 August 1862
DDW 71/862/16

In Sir Rodericks absence (he is in Bohemia & will probably be absent until the middle of October) your letter of 30th July with the enclosure from Mr. Joass[1] has been put into my hands.

I wish I could give you an opinion worth having on the tracks, but beyond feeling pretty sure that they were not made by any vertebrate animal, and that they were made by some Armeloid creature, I have not much to say about them.

The choice probably lies between Armelids & Crustacea.

I am very sorry to find that Mr. Duff's collection is not attainable.[2]

If I am in the neighbourhood of Elgin this year, as is not impossible, I shall ask to see it and then the Executors had better keep a sharp lookout, for if that tail is handy, my conscience will certainly give way & I shall steal it.

Notes

1. Joass was a Minister of the Church of Scotland on the north shore of the Moray Firth and a frequent companion of Gordon's. Referred to here are the tracks found at Tarbat.
2. Patrick Duff died in 1862.

To: Huxley
From: Birnie
Dated: 4 August 1862
DDW 71/862/18

I was glad to receive your report on the tracks at Tarbat.[1] Excepting the reptilian footprints, they are the only thing of the sort that I have seen in any of our Sandstones during my long inspection of them, and hence these Rosshire traces of animal life are valuable in my eyes.

I hope you have opened and read my last note (of Saturday) to Sir Roderick Murchison, as it gave you, of Jermyn Street, the last chance of getting Mr. Duff's cabinet.

If you have not already answered me by today's post I think it will be sold to Mr. Powrie, as I am not aware the Executors have any other offer, & you may yet get your tact (?) by fair means from him.

If you are this year to come north [of] the Grampians & the 'neighbourhood of Elgin' I rely upon your giving me some notice of your approach that I may be in the way & have the pleasure of seeing you here.

Note

1. These may be seen today in the Elgin Museum.

To: Gordon
From: Jermyn Street
Dated: 7 August 1862
DDW 71/862/19

I have opened your letter addressed to Sir Roderick but I suppose it is now too late to do anything – : nor indeed could I act in such a matter without Sir Roderick's express sanction.

I am glad if the specimens do not come to us that Mr. Powrie should have them.

He is a very good friend of mine and I shall try to get upon his weak side in order to extract the coveted tail from him.

If I am anywhere near Elgin I shall have much pleasure in reminding you of your kind invitation.

To: Gordon
From: Jermyn Street
Dated: 22 April 1863
DDW 71/863/40

Sir Roderick Murchison has just been talking to me about the slab of footprints mentioned in your letter to him, and he has asked me to request that you will be so good as to forward it *here*.

I wish the question of the age of the famous bed would get itself settled once for all & then I could bring out my monograph in peace.

To: Huxley
From: Birnie
Dated: 24 April 1863
DDW 71/863/44

Unfortunately the slab with footprints (rather casts) from Portmahomack[1] was sent off 1st rail from Elgin paid to Aberdeen, then per steamboat to London – a few hours before I got your letter of 22nd. It was addressed 'Sir R. M. Geol. Soc. Rooms Somerset House London'. As I intended it for him I have little doubt you will easily get it sent to Jermyn Street.

You have to thank my good & talented friend Mr. Joass for raising afresh the troubles (in which I glory in being a fermenter) that will not as yet allow the Moray Firth reptiles to be quietly set down as Triassic. Occasionally a scute or bone turns up at Lossiemouth; be assured did anything new come under my eyes I shall endeavour to send it forthwith to Jermyn Street.

Note

1. Where Joass, the amateur geologist referred to in this letter, was Minister.

To: Huxley
From: Birnie
Dated: 10 August 1866
Imperial College [draft in the Elgin Museum: DDW 71/866/45][1]

In consequence of my submitting a letter I had received from Sir Roderick Murchison, to my friend Mr. Grant of Lossiemouth, he (Mr. G) has kindly agreed to send the new specimen of *Telerpeton*(?)[1] and a bit of the *Hyperodapedon* for your examination.

As this is the season when you will possibly be leaving London for the country, you will be so good as to drop me a note mentioning at what time it will be most convenient for your taking up and finishing your inspection of these fossils; – as we are all wishful to have these rarities, after being vastly enhanced in value and in interest (by passing under your eye), safely back to the North again! For indeed we are beginning to suspect that the *fossil reptiles*, as it is alleged of *living bipeds*, seldom find their way home again once they cross the border.

Yea! more as to these fossils, they seem to be afraid to show themselves in the South, – as in 1864 & 5 I saw most of the geological collections of London, Dublin and Edinburgh, yet, of all the specimens we have sent from this quarter, one or two scutes of *Stagonolepis*, were all we could see open to the public eye; & they were in the Museum of the *Dublin University*.

Several visitors have expressed surprise that none of the Elgin reptilian remains & footprints were to be seen in the cases of the Museum in Jermyn Street. It raised the suspicion with some that our contributions were not valued. This however I know to be unfounded.[2]

Notes

1. This is the only known example of there having survived both the letter as sent and Gordon's draft. The Gordon letter is in the Huxley papers (Volume 17, n. 83) at Imperial College. Because this is a unique situation the draft is reproduced here in this note in its entirety, so that an assessment can be made of the extent to which Gordon made changes when he composed a letter with the previously written draft in front of him.

   Here is the complete text of the draft (i.e. DDW 71/866/45)

   In consequence of my submitting a letter I had from Sir Roderick Murchison Bart, to my friend Mr. Grant of Lossiemouth, he has kindly agreed to send the new species of *Telerpeton*(?) and a bit of *Hyperodapedon* for your examination. As this is the season when you will possibly be leaving London for the country, you will be so good as to drop me a note mentioning at what time it would be most convenient for your taking up and finishing your inspection of these fossils, as we are all wishful to have these rarities, after being vastly enhanced in value and in interest (by passing under your eye), safely back to the North, for we begin to suspect that the *fossil reptiles*, as it is alleged of living bipeds, seldom find their way home again once they cross the border. Yea more of the fossils – they seem to be afraid to show themselves [but in London] in 1864 & 5 I saw most of the Geological Museums collections of London, Dublin and Edinburgh, yet of all the specimens we have sent from this quarter, one or two scutes of *Stagonolepis* were all I could see open to the public eye. Several visitors expressed surprise that none of the Elgin reptilian remains & footprints were to be seen in the cases of the Jermyn Street Museums. It raised with some the suspicion that they were not valued. This, however, I knew to be unfounded.
2. The question mark is Gordon's.
3. The fact that only one letter, from all the letters Gordon wrote to Huxley, is now to be found in the Imperial College archive, is sufficient cause for scepticism about the comprehensiveness of that large and important collection. The real Huxley is not necessarily represented by it.

To: Gordon
From: 8 Beach Terrace, Littlehampton, Sussex
Dated: 18 August 1866[1]
DDW 71/869/19

I left town for the place where I shall be, saving the Association week, for the next six weeks, on Saturday so that your letter has been forwarded.

Pray express my obligations to Mr. Grant. I shall be very glad to find the specimens awaiting me in Jermyn St. on my return in the last week of September.

It is too bad my modesty & honesty should be turned against me! I have never exhibited any of the Elgin specimens in Jermyn Street because I thought that one of the lawful possessors thereto – 'steering' round the galleries – would have cried 'Here is a [ ] reptile lifting [ ] has cooly taken possession of our property!'

Tell me what you present to the Museum & you may depend upon it there shall be no lack of visibility on the part of the specimens in our cases.

The fact is that when the great guns of Geology began firing at one another over my head,[1] in their Devonian Trias showdown I thought that I & my four footed friends had better lie quiet until the storm was over – and that point of stratigraphy had got itself settled. I became a scientific Macawber and hoped something would turn up.

But nothing has turned up & I was only thinking the other day that I might as well wait no longer & put forth all my notes & plates which have been finished these two or three years.

Your reminder therefore especially as it comes in the form of new material is especially welcome.

Notes

1. Dated 18 August 1869 by the Elgin Museum but clearly part of the correspondence about Grant's specimen of *Telerpeton*.
2. This may be strictly speaking true, though Lyell had not altered his opinion since 1859, while Murchison was to correct his observations about the Old Red Sandstone in the 1867 edition of *Siluria*. As mentioned in Part One, Huxley had other reasons for delay.

To: Huxley
From: Birnie
Dated: October 1866
DDW 71/866/41

I have this day delivered to the Highland Railway Co. two boxes addressed to you in Jermyn Street.

In the larger box you will find several specimens (*Stagonolepis*) belonging to the same block.

I have gifted these specimens in Box 'A' to the Museum of Aberdeen University.[1] Be so good as to tear off the *upper* address and forward the whole when you are satisfied with them to the Granite City as the under address directs.

The whole of the specimens, in the smaller box, belong to Mr. James Grant, Genl. Assembly *Schoolmaster*, Lossiemouth, Elgin, and are from the reptilian beds at that fishing village. The box contains three apartments. In the largest are packed the *Telerpeton* remains, or rather moulds.[2] I think they will unfold still more distinctly (Particularly the head & cast of the brain with its lobes?) than the first specimen, which I once more saw the other day at Mr. Powrie's, the [ ] of which specimen was from the Spynie quarry.

In the middle sized apartment are Mr. Grant's specimens of *Stagonolepis*. They will allow you to describe the structure of the skull better than any former specimen, althou' it is to be regretted they are so fragmentary. The scutes and small bits of bone were found at the same time as the skull & were considered to belong to the same animal.

In the smallest apartment is a fragment which I regard as belonging to *Hyperodapedon*. It shows some of the pavemented teeth. Any further [ ] [the draft breaks off at this point]

Notes

1. Presumably these are the specimens now in the Museum of the Department of Zoology at Aberdeen.
2. It is this material that carries Huxley forward by allowing him to write 'On a New Specimen of Telerpeton Elginense' (see Appendix III). Evidently, however, Gordon did not restrict himself to having only one set of moulds or casts made. George Campbell acknowledged the receipt of a set from Gordon on 24 January 1867.

To: Gordon
From: Jermyn Street
Dated: 27 October 1866[1]
DDW 71/869/23

I am back in full work and quite ready to give you a report upon the *Telerpeton* whenever you are so good as to send him up.

Note

1. This note is numbered DDW 71/869/23 but the date is more than usually difficult to read. On 27 November 1867 Huxley refers to the 'specimen of *Stagonolepis* which came along with Mr. Grant's *Telerpeton*'. Huxley's paper was read at a meeting of the Geological Society on 19 December 1866. This note must have been written after DDW 71/866/41 but before Huxley's talk.

To: Gordon
From: The Athenaeum
Dated: 12 November [1866]
[American Philosophical Society Library (B/H 981.30)]

I write a hasty note to acknowledge the safe arrival of the fossils. They came late on Saturday and as today is one of my two lecture days I have had no time but to take a glimpse of them. But the *Telerpeton* promises to be highly interesting.

To: Gordon
From: The Geological Survey of Great Britain
Dated: 22 November 1866
DDW 71/866/45

I am sure you qill be glad to hear that the *Telerpeton* has turned out to be a regular jewel. I have been working it out very carefully in such spare time as a host of lectures & examinations would permit – and I have made out almost as much of it as if it were a recent skeleton.

I have episcapated (I know that is a good Scotch word) the forefoot and the entire hind foot,[1] with a good deal more of the tail than was visible when it reached me, & the entire lower jaw with two tusks in front of it. My work went on in fear & trembling, but I am happy to say I have had no mishaps. The beauty of the whole affair is that *Telerpeton* is a lizard of an altogether modern type & much more like some that run about now, than it is to the extinct mesozoic or palaeozoic Saurians. Is Mr. Grant open to an offer for the specimen?

The fragments of *Stagonolepis* promise to be most instructive, but I have not taken a good look at them yet.

Note

1. Of which Huxley provides a drawing in 'On a New Specimen of *Telerpeton Elginense*'.

To: Huxley
From: Birnie
Dated: 16 August [1867]
DDW 71/867/16

On the present occasion I do not intimate the approach of new specimens but the departing of old. It is intended under the auspices of Mr. Powrie to exhibit at the Dundee meeting as many as possible of the interesting geological specimens collected in the North of Scotland. As some of the most interesting are at present in your keeping, I have to request that you will kindly transmit to the 'local secretaries of the Brit. Ass. at Dundee' – Mr. Grant's specimen of the *Telerpeton* that of the *Hyperodapedon*, the specimen of the cast of small scutes of *Stagonolepis* which was sent to Jermyn Street for inspection [illegible additions at this point] and from which Agassiz inferred that the animal was a fish, together with any other specimens that may have been given to yourself or presented to the Jermyn St. Museum. This would be a good opportunity of transmitting so far north, the large specimen (scutes of *Stagonolepis*) which as I wrote to you at the time I had made over to the Museum of Alma Mater – 'Marischal College, Aberdeen'.[1]

To do this will be a bother to you, whose valuable time is so well occupied, but I venture to hold out as an inducement a promise that any fresh specimen from Lossiemouth or elsewhere in this locality coming into my hands, will be sent for inspection in the first instance to Jermyn St.

Mr. Grant has not sold & I question if he now intends to sell the *Telerpeton* [ ]

[The final sentence was written across the already used page and the last part of it is impossible to decipher.]

Note

1. See letter of October 1866 (DDW 71/866/41). Huxley seems not to have answered Gordon on this point. See next letter.

To: Gordon
From: Geological Survey of Great Britain
Dated: 30 September 1867
DDW 71/867/20

I left London on 15th August & did not return here till the day before yesterday. No letters were forwarded to me as to secure peace & quietness I had voted myself out of reach by the post.

Hence I knew nothing of your letters till Saturday and even if the first had reached me I fear I could have done nothing, as the place shuts up every year from the tenth of August till the last of September.

However, I am very sorry your wishes were not complied with, and I hope you will forgive my seeming neglect of your letters.

I greatly wished to visit Dundee & the north of Scotland for several reasons; among the rest because I had an invitation from Mr. Grant Duff to visit him & meet you in Elgin.

But in the summer I became too much in need of rest, that I slipped out of all my engagements.[1]

I think I must write to Mr. Grant to see if I cannot tempt him to let us have the *Telerpeton*. Many thanks for your intermediation with him. I shall try to get out my monograph about the Elgin fossils this winter as it seems useless to wait for new materials.[2]

Notes

1. This is the stress of work that resulted in breakdown in the early 1870s.

2. As already mentioned, Huxley's position had been greatly strengthened by the discovery of *Telerpeton* and *Stagonolepis* remains in the same strata. His waiting for new materials, however, confirms his disinclination to be a field geologist.

To: Gordon
From: [London]
Dated: 1 November 1867
DDW 71/867/23

Acting on the suggestion contained in your last letter, I requested Sir R. Murchison to make an application to the Council of the Royal Society (of which I am just now a member) for a grant from the Donation Fund to enable you to make any excavations that may seem good to you in the Elgin Sandstones, and yesterday we voted you £50 for that purpose. So now I hope you will 'go in and win' for the importance of the Elgin fossils waxes greater every day.

The other day some fossils from Central India were placed in my hands & I at once recognized among them [ ] of *Hyperodapedon*.[1] I should never have been able to make this identification except for the unlikely looking specimen you sent me up last summer. Now there is reason to believe that these Indian beds are of like age with those from which I got *Dicynodia* a year or two ago – and *Dicynodia* is as you know the characteristic reptile of the great South African sandstone formations, so that I am in great hope of being able to get one great horizon from the Manse at Birnie to pretty near the pagodas of Juggernauth – & perhaps to Australia. But for the purpose I want some more from Elgin, and especially a skull of *Stagonolepis*.

Notes
1. These were the fossils that had been in the rooms of the Geological Society for a number of years, something it was not necessary to tell Gordon at this juncture.

To: Huxley
From: Birnie
Dated: 2 November 1867
DDW 71/867/24

I have just received your letter of yesterday, announcing that Sir R. M. had applied for & obtained a grant of £50 from the Royal Society for the purpose of obtaining reptilian remains from the Elgin sandstones. I am gratified by having the expenditure of the sum entrusted to my hands; & I have in consequence already written to my friend Mr. Grant who is on the spot to inquire as to the existence of some scutes (etc.) that were discovered some time ago in a part of the rock which the workmen thought too friable for their purpose & left completely untouched. The present season of the year, the short day & cold of winter, must cramp our energies; but I do expect, ere the summer is far gone, to be able to send you, if not your *desideratum* (a head of *Stagonolepis*) at least something additionally illustrative of our Reptiles which in days of old must have had cousins scattered over the face of the earth as widely as Queen Victoria's subjects are at this day.

To: Gordon
From: Jermyn Street
Dated: 27 November 1867
DDW 71/867/28

I ought to have written to you before this to say that I have by no means overlooked the specimen of *Stagonolepis* which came along with Mr. Grants *Telerpeton* & which I thought had been sent by yourself. It is undoubtedly part of the snout of *Stagonolepis* and tells an interesting story – but it is not enough. And I am sure when you get to work that it will be superseded by that complete skull which I see looming in the future, with prophetic insight.

By the way is Mr. Grant still coy about parting with *Telerpeton*?

I do not like to bother him about it, but I break the tenth commandment whenever I look at it.

I hope the sin is put down to my official & not my private account.

To: Gordon
From: The Geological Survey of Great Britain
Dated: 7 January 1869
DDW 71/869/17

My paper on *Hyperodapedon Gordoni* & its Indian & South British relatives is coming off on Wednesday next (12th Jan).[1]

Have your excavations during the past year brought anything anew to light? I shall be very glad to be informed if they have.

A friend of yours I understand left a cast here in my absence. It is hard to interpret a cast safely but it looks like a big fish jaw, such as is found both in the Old Red & in Triassic formations.

Did I tell you that in addition to the Warwickshire specimens, *Hyperodapedon* has now been found in the Trias of Devonshire & in Central India.[2]

Notes

1. 'On *Hyperodapedon*', *Quarterly Journal of the Geological Society* (1869): 138–152, read at a meeting of the Society on 13 January 1869.
2. These are all referred to in Huxley's article. It would appear that he had not previously told Gordon about them.

To: Huxley
From: Birnie
Dated: 11 January 1869
DDW 71/869/2

Absence from home has prevented an earlier reply to your note of the 7. I regret I have to report that no reptilian remains , save single scutes, have last year turned up in our sandstone quarries, so far as I or any of my friends have seen or heard of. Save a few pounds which I payed out indemnifully (?) at Findrassie, the whole of the Royal Society grant lies in the bank, ready for extracting & seeing any specimen that may show itself to the quarryman. We must wait such an appearance, for indiscriminate quarrying for these fossils would be expensive & [ ].

Now that your paper on *Hyperodapedon* is completed, I do hope that the reptilian remains sent for examination to Jermyn Street by Mr. Grant,[1] & the directors of the Elgin Museum, will be returned without further delay. I pledged my word that they should be. I tell you plainly that until this be done they will not send any more specimens that now or may afterwards fall into *their* hands. Of course you will duly receive for the Royal Society any fossils that I can secure through the grant sent me by that learned body.

It is likely that the cast left with you is the one which my friend Dr. Boyd got from me some time ago. The original is in the Elgin Museum. Little could be made of it *per se*; but we are collecting as many fragments as possible from the beds of the undoubted Old Red, in the hope that by and by something may be made of them. One specimen, part of a large [ ] (much larger than I ever saw from the [ ] beds) unmistakeably shows it had the same ornamentation in its plates as that which have hitherto got the name of [ ]. Among the specimens already collected there are some of the same sort as those which I recollect your looking at in the Geological Society. They were given to him to be reported on, but what has become of them I know not, and this has been another reason why the Directors of the Elgin Museum cast a veto on any more of their specimens being sent to the omniverous Metropolis. Your returning the reptilian specimens will help me in getting this veto removed.

[after the signature]

When in Nairnshire last week, I learned that a piece of Sandstone with some form [   ] appearance, had been sent to the Geol. Soc. in London. I did not see it.

Notes

1. How to get Huxley to return specimens to Elgin had by this time become a major problem for Gordon and his friends in Moray. Evidently Gordon sent copies of this letter to some of these friends, since on 13 April James Grant wrote to him from Lossiemouth saying:

   Your answer to Professor Huxley is precisely to the point. I was glad to see you had given him a hint as to the restitution of the fossils. There is no doubt whatever, as you stated, but there will be a shyness by and by in sending fossils to him for examination should a difficulty be anticipated in getting them back. Now that he has examined the *Hyperodapedon* and *Telerpeton* I would suppose he would return them without delay. I wonder if he ever found the head of *Stagonolepis* I sent along with the *Telerpeton* of any use to him [DDW 71/869/3]

   Perhaps unwittingly Huxley killed the goose that laid the golden egg. He had never been in a position to judge the thoroughness of on-site field work that resulted in his being sent specimens. Now this situation was exacerbated by his failure, perhaps because of pressure of work, but perhaps partly because of a proprietary attitude to the fossils he had in hand, to deal fairly with those who had generously shared their finds with him. From this point on, he wouldn't automatically be told about finds that related to his published work. It was not, incidentally, until 1877 that he corrected his earlier error about the head of *Stagonolepis*, no doubt on the basis of the specimen sent to him by James Grant.

To: Huxley
From: Birnie
Dated: 17 November 1869
DDW 71/869/27

I was at Lossiemouth yesterday, and (from a large mass, some tons in weight, of blocks, all containing scutes or bones which however seem unfortunately to have been separated & scattered before they were fixed in the rock) I selected one, in two pieces which evidently contains teeth of *Stagonolepis*, & which I hope will reach you in a day or two, as now despatched 1st rail & *paid for*. In the box you will find some sulphure (?)

of iron, which I think have had their origin from some organic matter. Do their *shapes* indicate anything? – as coprolite for all trace of *organic structures* seem to have been destroyed, if this can [ ] by the crystalization.[1]

Note

1. Not one of Gordon's more distinguished drafts, this one literally tails off into incoherence.

To: Gordon
From: Geological Survey of England and Wales
Dated: 28 November 1869
DDW 71/869/31

I was overwhelmed with work all last week and could not find time to examine your last package. (By the way don't take the trouble to pay carriage hereafter as all that is properly official). There is a fine series of scales in the big block and the bit of upper jaw is invaluable. I am having the big block worked at to see what can be found in it.

By the way, your packer did not do his work very well & the blocks got terribly knocked about & shattered in transit.

To: Huxley
From: Birnie
Dated: 31 March 1870
DDW 71/870/2

During the winter nothing has been added to the reptiliferous blocks collected for you last autumn at Lossiemouth. As some of the apartments are filled I get them cleaned out for future . . . [indecipherable phrase][1]

In October when I had the pleasure of seeing you in London, you had the idea of running down at Easter to examine them, rather than to have them sent to you in Jermyn Street. There is a large collection of them with at least a part of bone or scute (of *Stagonolepis*?) in each. These bones & scutes are so shattered that most of the specimens may be of little value. But I would not venture to make a selection of them. Perhaps, if you find it impossible to come and personally inspect them you may get some palaeontologist friend to run north to see the larger blocks broken up & pull up out from the whole what was worth your closer study.

Note

1. This is not the only time when Gordon shows himself to be thinking his way through a subject while writing the draft.

To: Gordon
From: Geological Survey of England and Wales
Dated: 14 April 1870
DDW 71/870/3

I have put off replying to your note in the hope that I might be able to get to Elgin this Easter.

But I am so terribly tired & used-up with a very hard winters work that I am recommended to get a week or ten days of complete change away from fossils & all their works – and so I am going to run away to the Rhine country tonight & have a tramp there with my old friend Dr. Hooker.[1]

I shall try to get to Elgin in the course of the summer as I want very much to see what you are doing. If the British Association meets in Scotland in 1871 as I think is very likely we will have a regular Elgin day all to ourselves as Section C.

Note

1. This is Sir J. D. Hooker, later President of the Royal Society. Of the expedition or holiday referred to here, Leonard Huxley simply says, 'From the 14th to the 24th of April Huxley, accompanied by his friend Hooker, made a trip to the Eiffel country. His sketch book is full of rapid sketches of the country, many of them geological . . . ' (*Life* I, 329). In as far as Leonard Huxley had the task of making sense of the T. H. Huxley papers in his possession and of using the letter texts to weave an account of a long life, it would be ungenerous to carp about the fact that he wrote the biography without doing much original research. Nonetheless, what Hooker and Huxley had in common as scientists remains to de described and discussed.

To: Huxley
From: Birnie
Dated: 14 July 1871
DDW 71/871/17

I trust that before you return from Scotland next month that you will arrange to see the blocks of reptiliferous sandstone excavated at the expense of the Royal Society and now lying at Lossiemouth. In each there is some *bit of bone* or scute; but none of them promise much more even when broken up. If Mahomet will not come to the mountain the alternative is that the mountain must move. How then would it suit you as the simpler plan were I to send you a truckful of the best of them to Edin. during the Assn. Week?

Perhaps you would there find time (at intervals) to examine them. I intend going to Edin. in the first part of next month; but, from some engagements will not be able to return home immediately on the closing of the Association. However, Mr. Grant of Lossiemouth could act as your guide. I wish you would bring with you to Edin. his small specimen in which there are two Telerpetons & which I left with you about a couple of years ago. He is always on the spot, is likely to get the first intimation of any good thing & wishes back one specimen before he sends another.

To: Gordon
From: Geological Survey of England and Wales
Dated 22 July 1871
DDW 71/871/18

I expected a scolding when I opened your letter but you have had mercy on a poor man who has been fairly overwhelmed with schoolboards[1] & other worriments for the past year. I am very much obliged for your kind offer to send some of your blocks to Edinburgh, but I have always found any serious work so difficult in the hurly burly of the association – that I do not wish to put you to that trouble. I suppose the best way will be for 'Mohamet to come to the mountain', when you are back in the Manse. I am going to give myself a serious holiday this year – and to that end have taken a house at St. Andrews for the months of August & September. Now St. Andrews is 'quite convenient' as the Irish say, to Elgin and when we meet as I hope we shall, at Edinburgh I will arrange with you when it will be best to go & look at the stones *in situ*.

Moreover I positively am going to send back a whole lot of specimens. I had them looked out before your letter arrived. Such is the force of conscience.[2]

Mr. Grant's specimen shall be seen to.

Notes

1. Leonard Huxley enumerated some of these worriments: the lectures to working men, postponed from October 1870 to February 1871; and public lectures in Leicester, Birmingham and Bradford. Huxley was elected to represent the Marylebone division of the newly formed London School Board, which first met on 15 December 1870. Having once committed himself, he gave the Board a lot of his time and energy until early in 1872, when he resigned 'on account of his health and private affairs'.
2. Conscience shaken or not, he did not do it in 1871.

To: Gordon
From: Geological Survey of England and Wales
Dated: 28 November 1871
DDW 71/871/32

The 'muckle stanes' have arrived & as far as I have been able to see with no great damage to the contents. But the convenience of examining them here is less than it was in your coal shed.

The other day I recollected my promise to write to Sir H. James. I value his answer as I think the best way would be for you to write to him direct saying exactly what maps you want.

To: Sir H. James
From: Gordon
Dated: 29 November 1871
DDW 71/871/32[1]

In a letter just received from Professor Huxley, he informs me that you have very kindly promised to send me such sheets of the map of Elginshire as would enable me to mark for him the localities where remains of reptiles have been found in our upper red Sandstones. For this purpose the three (1, 2 & 7) I have tinted on the accompanying 'Index to Elginshire' will be sufficient & the six inch scale the most suitable size.

Note
1. Written on the back of the previous letter.

To: Huxley
From: Gordon
Dated: 5 July 1874
DDW 71/874/2[1]

The six & the one inch Ordnance maps of our Reptilian region are at length printed, & Sir Henry James has kindly sent me early copies.

You wished me to mark on them some particulars relating to the locality of these fossils. I shall be happy to do so if you [will] be so good as to tell me what particulars you desire beyond defining the place where the footprints are seen & where the fossil bones & scutes or their moulds (as at Findrassie) have been found.

Nothing new has turned up. I have still £30 in deposit at the [bank?] of the £50 the Royal Society gave me. It will be drawn upon as soon as any good specimen indicates its appearance in the quarries at Lossiemouth.

Perhaps you recollect some fragmentary fossils from the undoubted Old Red which I offered you for examination; but you said your hands were full. I know neither to whom they should be sent to test my idea that they will prove that specimens we in the North were wont to call *Bethendipus* were parts of a large *Pteridithus* (?)

Note

1. Exceptionally, this draft was written on the back of a special notice of the sale of shares by a London banker in 'The Industrial Coal and Iron Company, Ltd'.

To: Gordon
From: 4 Marlborough Place
Dated: 25 July 1874[1]
DDW 71/874/3

Letters addressed to Jermyn St. do not reach me very regularly just now as my daily work lies, for the present at South Kensington.[1]

I shall be greatly obliged if you will mark down on the 6 inch maps not only the localities in which the fossils & footprints have been observed, but anything else bearing on the age of the deposits you may think desirable. No one knows so much about this interesting region as you do.

I am sorry that my 'hands are still full' – all the fuller for a year and a half's illness,[2] which has kept me in a wretched condition of incapacity. However, I have happily shaked all that off and have fair hopes of overtaking my work in course of time.

Why not send your fossils to Dr. Tracquair[3] the newly appointed Curator of the Edinburgh Museum. He is a very sound man & learned in fossil fishes.

As I am coming North at the end of February to do my Rectorial duties at Aberdeen but I fear I shall not be able to fetch (*sic*) Elgin.

Notes

1. This letter is dated 1876 in the Elgin catalogue, but Huxley's 6 is difficult to distinguish from his 4. The rectorial address referred to in the letter was delivered in 1874.
2. Because the biology department of the Royal School of Miners in Jermyn

Street (whose building, incidentally, did not survive the Second World War) had by this time been transferred to the new South Kensington Museum.

3. The gap in the Huxley-Gordon correspondence between November 1871 and January 1874 corresponds, of course, to the period of relative incapacity. Huxley had spent three months in Egypt but had not recovered his health there. A strict diet improved matters but it is probably true to say that he never fully recovered his health.
4. This probably marks the end of Huxley's interest in Elgin, except for the production of the monograph of 1877, the drawings for which had already been prepared. In a letter to George Gordon dated 6 November 1891, written on behalf of E. T. Newton, (DDW 71/891/36) the latter reported that Tracquair 'had a good sharp lesson in returning specimens'. The Director of the Museum had taken stern measures and persuaded him to return a thousand! Since Tracquair's appointment, work on the Elgin fossils had been done in Scotland, largely in conjunction with the Geological Survey.

To: Huxley
From: Birnie
Dated: August 1874
DDW 71/874/11

I wish to send you a block from the Reptilian Sandstone which contains some organisms that may interest you. The bones indeed look as if they had been very much deranged before they were embedded; but I think the specimen is complete, that is not a part of a larger animal. It appears to me to have a head & a vertebral column, but I must not speculate. I have procured it and boxed it at the expense of the Roy. S. of whose grant to me still remains in my hands, & I propose to send it to their Secretary if you do not give me order for any other address.

To: Gordon
From: Athenaeum Club, Pall Mall
Dated: 8 August 1874[1]
DDW 71/874/10

Your letter [ ] come into my hands.

I shall welcome the reptile 'fra' the north and it is just as well you should send it to the Royal Society as the Museum in Jermyn St. will be shut up for the next month.

I trust that you are well and vigorous as ever.

Note

1. This must be the reply to DDW 71/874/11, though the two letters have been catalogued in reverse order.

To: Gordon
From: Edgehill, Culter, Aberdeen
Dated: 23 September 1875
DDW 71/875/34

I am coming to Elgin to have a final look at the *Stagonolepis* remains in the museum before publishing my monograph which has been so long in hand. I shall reach Elgin next Monday (27th Sept) and shall put up at the Hotel for two or three days. I suppose I shall have no difficulty in getting a bed room & sitting room there.

Will you have the kindness to give directions at the Museum, so that I may have access to the specimens? You see I have a lively recollection of the stern janitor.[1]

Note

1. The stern janitor who had impeded Huxley's entry to the Museum on the only other occasion he had visited it may have been William Ingram, the first keeper. John Martin (1839–1881) was the first curator.

To: Huxley
From: Birnie
Dated: 22 March 1878[1]
DDW 71/874/4

Last autumn, when Professors Nicol (of Aberdeen) and Heddle (St Andrews) were here on a mineralogical excursion, I took them to the *Findrassie* quarry. Professor Nicol found a slab with the impressions of two bones & of some scutes (of *Stagonolepis*) which he has now in the Aberdeen Museum.

On narrowly examining the rock, the layer or part of the bed in which the fossil remains seem to lie, was detected pretty low down in the escarpment.

They've since got some tons of rock quarried, and, although some vertebrae, bone & scutes have been got, yet no trace of skull or jaw has been seen.

Addressed to you in Kensington, I have forwarded, per steam boat from Aberdeen, two large slabs (which lay, the one on the other, separating in the plane of the scutes etc), and every smaller bit of rock in which I could detect anything beyond a scute. There may be something better in the interior than what appears on the surface of the larger blocks. At all events, these specimens of this rare reptile, if superfluous for the Kensington Museum, will be acceptable gifts to any other collection. They are all parts of the fruits of the Royal Society grant, of which some £20 still remain in my hand.

Some appearances have been lately observed which tend not a little to support my view viz. that it is difficult, if possible, to *separate our reptiliferous beds from those of the upper beds of the Old Red Sandstone* where *Holoptychian* are found.

1. In the Bishopsmill beds, where *Holoptychian* scales have been got, there have been found two slabs with markings, which I am almost persuaded are reptilian footprints, large like some from Cummingston.
2. Mr. Linn, of the Geological Survey, has found undoubted Old Red Sandstone fish remains in rocks lying to the west of Stotfield (Lossiemouth) – a district where all has hitherto been considered reptiliferous.

Professor Geikie, his men tell me, is expected here next month,[1] & we shall see what light he will throw on the *Questio vexatio*.

In the block sent, you will observe black streaks & other foreign matter. What are they?

P.S. Accept my best thanks for the elaborate monograph on *Stagonolepis*, for which I am no doubt indebted to you.

On the edge of the larger slab & at the mask of the 'red cross', I observe something like the cast of a jaw with *teeth*, but far too small for the tooth of a *Stagonolepis* so large as the scutes & vertebrae indicate.

Notes

1. This letter provides a clear example of Gordon having added the date on some subsequent occasion. Gordon, however, got the date wrong, since he acknowledges here the receipt of Huxley's monograph.
2. Possibly the specimens now in the Museum of the Department of Zoology at Aberdeen.
3. As Director of the Geological Survey of Scotland. See R. B. Wilson, *A History of the Geological Survey in Scotland*, National Environment Research Council and Institute of Geological Sciences, 1977.

Appendix I
Select List of George Gordon's Publications

A THE SCIENTIFIC PAPERS

Included in Part A are those articles Professor Harvie-Brown had gathered together when Gordon's son asked him to prepare them for publication. The bound volume which contains them in the Elgin Museum, which is the principal memorial to Gordon's scientific work, includes duplicates, off-prints, corrected off-prints and interleaved autograph drafts. They are given pride of place in this listing of Gordon's publications because they had been identified as his most important work, although they were never published as a volume.

1. 'A Fauna of Moray', *The Zoologist*, Volume II, February 1844, pp. 421–28, 502–515, 551
   This was originally a work copy and is interleaved. Bound in is a second copy of this article with notes by Charles St. John Eyre of Inverness.
2. 'A List of the Fishes that have been Found in the Moray Firth and in the Fresh Waters of the Province of Moray', *The Zoologist*, Volume X, May-June 1852, pp. 3454–62, 3480–9.
3. 'A List of Crustaceans in the Moray Firth', *The Zoologist*, Volume X, December 1852, pp. 3678–87.
4. 'A List of the Mollusca Hitherto Found in the Province of Moray and in the Moray Firth', *The Zoologist*, Volume XII, 1854, pp. 4300–18, 4421–35, 4453–62.
5. 'List of Echinodermata Hitherto Met With in the Moray Firth', *The Zoologist*, Volume XI, February 1853, pp. 3781–85.
6. 'On the Hymenoptera (but chiefly the Genus Bombus) of the Province of Moray', *The Scottish Naturalist* [1852?].
7. 'On the Geology of the Lower and Northern Part of the Province of Moray; its history, Present State of Inquiry, and Points for Future Examination', *Edinburgh New Philosophical Journal*, New Series, Volume IX, January 1859, pp. 14–58.
8. 'List of Lepidoptera Hitherto Found Within the Province of Moray etc.; Arranged According to Doubleday's List, Second

Edition', *The Zoologist*, Volume XIX, August 1861, pp. 7763–75.

9. Wilson's Prehistoric Animals, read with reference to the Archaeology of the Province of Moray'.
A paper read to the Elgin Literary and Scientific Association on Tuesday, 2 February 1864 and printed by James Black in Elgin.

10. *Stromness (Orkney) as a Summer Residence*, James Watson: Elgin, 1885.

11. 'Notice of a Hoard of Silver Coins Discovered in Banffshire, of which Three are now Presented to the Museum'.

12. 'Obituary Notice of Dr. Hugh Falconer', *Proceedings of the Royal Society*, Volume XV.

13. 'Hints for obtaining the Finest Honey without Destroying or Starving a Single Bee.

B SHORT NOTES IN *THE ZOOLOGIST*

The Zoologist A Popular Miscellany of Natural History was 'conducted' by Edward Newman and published in London by John Van Voorst. It was issued in monthly numbers and annual bound volumes. The preface to the first volume in 1843 includes the disarming statement: 'Everyone who subscribes a single fact is welcome – nay, more than that – has a direct claim to be admitted as a contributor.'

1844	II, 474–5	'On the Scarcity of Wasps in 1843'.
	II, 653–4	'Summer Birds in Elgin in 1844'.
1845	III, 807–8	'Note on the Habits of the Honey Bee'.
1846	IV, 1521	'Occurrence of the Locust in Scotland'.
1847	V, 671	'Preserving the Honey Bee'.
1848	VI, 1970–1	'Breeding of the Salmon'.
	VI, 2008	'Habits of the Field-Mouse'.
	VI, 2297–98	'Prolific Pair of Blackbirds'.
	VI, 2304	'Answer Respecting *Larus Maximus*' .
	VI, 2326–8	'Suggestions for Obtaining Crustacea'.
1850	VIII, 2763–4	'Variety of the Common or House Mouse'.
	VIII, 2769–70	'Waxwing in Various Localities in Scotland'.
1852	X, 3660	'Occurrence of *Regalecus Glesne* at Cromarty'.
	X, 3712	'Note on 'Sporting by Steam', Singular Capture of the Porbeagle Shark'.
1853	XI, 4100–2	'Note on a Variety of the Plaice'.
	XI, 4106–7	'Caprella Laevis in the Moray Firth'.
1854	XII, 4171–2	'Capture of the Three-Spotted Wrasse (*Labrus Trimaculatus*) in the Moray Firth'.
1857	XV, 5537–8	'Pupa Hunting'.
	XV, 5787	'The Locust in Scotland'.
1860	XVIII, 6860	'The Great Gray Shrike at Forres'.

1861	XIX, 7433–4	'Occurrence of the Little Bustard (*Otis Tetrax*) in Moray'.
	XIX, 7705–6	'Bats in Aberdeenshire'.
1864	XXII, 8889	'Sand Grouse near the Moray Firth'.
	XXII, 9360	'Redbacked Shrike, or Butcher Bird, in Forfarshire'.
1865	XXIII, 9457–8	'Adders in Lauderdale'.

C MISCELLANEOUS PUBLICATIONS

1. A Description of Birnie for the *New Statistical Account*.
2. An Introduction to Sir Thomas Dick Lander's *Account of the Great Moray Floods* (McGillivray: Elgin, 1873).
3. 'Ulleriore, of Spynie Loch, Elginshire', *The Scottish Naturalist* Volume III (1875–76): 127.
4. 'Phromina Sedentaria and Ihr Beroe', *The Scottish Naturalist* Volume VI (1881–82): 56–9.
5. 'The Victor Overthrown', *The Scottish Naturalist*, Volume VI (1881–82): 288.

No attempt has yet been made to gather together Gordon's many contributions to local newspapers.

Appendix II

A SELECT LIST OF ARTICLES BY T. H. HUXLEY RELATING TO HIS WORK ON REPTILIAN FOSSILS IN THE MORAY REGION

A convenient listing of Huxley's articles and books is to be found in Mario di Gregorio's *T. H. Huxley's Place in Natural Science*, (Yale University Press: New Haven and London), 1984, pp. 201–248. The select list given below is intended to chart the progress of the work being discussed in this book, even though to separate these articles from the flow of work in general is necessarily artificial. All these, and most other scientific papers by Huxley, are to be found in the four-volume (with supplement) Memorial Edition, entitled *The Scientific Memoirs of Thomas Henry Huxley*, (London: Macmillan, 1898). References to this edition in the list which follows are coded as *M*I, *M*II, *M*III or *M*IV.

1856 'On the Method of Palaeontology', *Annual Magazine of Natural History*, 18, (1856): 43–54. *M*I: 432–44.

1858 'On a New Species of Pleiosaurus (*P. Etheridgii*) from Street, near Glastonbury; with Remarks on the Structure of the Atlas and Axis Vertebrae and of the Cranium of that Genus', *Quarterly Journal of the Geological Society*, 14 (1858): 281–94. *M*I: 522–37.

1857–9 'On the Theory of the Vertebrate Skull', *Proceedings of the Royal Society*, 9 (1857–59): 381–457. *M*I: 538–606.

1859 'On the Stagonolepis Robertsoni (Agassiz) of the Elgin Sandstones; and on the Recently Discovered Footprints in the Sandstones of Cummingstone'. *Quarterly Journal of the Geological Society*, 15, (1959): 440–60. *M*II: 94–118.

1859 'On the Persistent Types of Animal Life', *Proceedings of the Royal Institution*, 3 (1858–62): 151–53. *M*II: 90–93.

1859 'On Some Amphibian and Reptilian Remains from South Africa and Australia', *Quarterly Journal of the Geological Society*, 15 (1859): 642–49. *M*II: 120–29.

1859 'On a New Species of Dicynodon (*D. Murrayi*) from near Colesburg, South Africa; and on the Structure of the Skull in Dicynodonts', *Quarterly Journal of the Geological Society*, 15 (1859): 649–58. *M*II: 130–40.

1859 'On the Dermal Armour of Crocodilus Hastingsiae', *Quarterly Journal of the Geological Society*, 15 (1859): 678–80. *M*II: 157–65.

1859 'On a Fragment of a Lower Jaw of a Large Labyrinthodont from Cubbington', *Memoirs of the Geological Survey of Great Britain*, (1859): 56–57. *M*II: 269–70.

1860 'On the Dermal Armour of Jacare and Caiman, with Notes on the Specific and Generic Characters of Recent Crocodilia', *Journal of the Linnaean Society*, 4 (1860): 1–28.

1861 'Preliminary Essay upon the Systematic Arrangement of the Fishes of the Devonian Epoch', *Memoirs of the Geological Survey of Great Britain. Figures and Descriptions Illustrative of the Fishes of the Devonian Epoch*, 2: 421–71.

1862 'Geological Contemporaneity and Persistent Types of Life' (The Anniversary Address to the Geological Society), *Quarterly Journal of the Geological Society*, 18 (1862): xi-liv. *M*II: 512–29.

1862 'On the Methods and Results of Palaeontology', *Medical Times Gazette*, 24 (1860): 363.

1867 'On a New Specimen of Telerpeton Elginense', *Quarterly Journal of the Geological Society*, 23, (1867): 77–84 (read to the Society on 19 December 1866). *M*III: 205–213.

1868 'On *Saurosternon Bainii* and *Pristerodon McKayi*, Two New Fossil Lacertilian Reptiles from South Africa', *Geological Magazine*, Volume V (1868): 201–205. *M*III: 298–302.

1869 'On Hyperodapedon', *Quarterly Journal of the Geological Society*, 25 (1869): 138–52, and the discussion, 157–8. *M*II: 374–90.

1870 'On the Pedigree of the Horse', *Proceedings of the Royal Institution*, 6 (1870): 129 (report).

1870 'The Anniversary Address of the President', *Quarterly Journal of the Geological Society*, Volume XXVI (1870): xxix-lxiv. *M*III: 510–50.

1875 'On Stagonolepis Robertsoni, and on the Evolution of the Crocodilia', *Quarterly Journal of the Geological Society*, 31 (1875): 423–38. *M*II: 68–83.

1877 'The Crocodilian Remains Found in the Elgin Sandstones, with Remarks on the Ichnites of Cummingstone', *Memoirs of the Geological Survey of Great Britain*, Monograph 3. *M*IV: 188–241.

1880 'The Coming of Age of the Origin of Species', *Nature*, Volume XXII: 1–4. *M*IV: 395–403.

1881 'The Rise and Progress of Palaeontology', *Nature*, 8 (September 1881), XXIV: 452–55. *M*IV: 508–19. *Collected Essays*, 4: 24–45.

1887 'Further Observations upon Hyperodapedon Gordoni', *Quarterly Journal of the Geological Society*, 43 (1887): 675–94. *M*IV: 636–57.

Appendix III

When Huxley obtained from Gordon in February 1877 a 'Geological Sketch of the District of Elginshire' (see Plate 3), Gordon added a note on the geology of the region, which Huxley included in Monograph III of the Memoirs of the Geological Survey of the United Kingdom. The full text of this note is given below.

NOTE UPON THE GEOLOGICAL STRUCTURE OF THE ELGIN DISTRICT, BY THE REV. GEORGE GORDON, D.D.

The lower or northern part of Elginshire, in which reptiliferous beds appear, has been visited and described by several geologists. The different papers, containing the result of their observations, are referred to by Sir Roderick Murchison in the *Quarterly Journal of the Geological Society* for August 1859, and by J. W. Judd, Esq., in the same Journal for May 1873.

For some time after the discovery of the *Telerpeton*, at Spynie, and up to the finding of a slab at Findrassie, which clearly proved the reptilian character of *Stagonolepis*, these beds, conformably following, and lithologically resembling the holoptychian, or undoubted "Old Red" strata, were supposed to belong to the same geological epoch. The next discovery of the *Hyperodapedon*, then induced most geologists to regard them as of a later, or the Triassic age.

Although the quarrymen are generally alive to any organism that may turn up in their labours, yet it is singular that no other remains (animal or vegetable) have as yet been detected in these reptiliferous beds; bones, scutes, and footprints forming the whole of their vestiges. There are concretions of pyrites, probably coprolites, but no organic structure has been seen in them.

The accompanying sketch is intended to mark the different localities where the various rocks come to the surface, and to point out their dip, where it has been ascertained. No outline, to show their probable range, has been attempted.

The district is so covered by drift, that no junction of the holoptychian and reptiliferous strata has been laid bare.

It is supposed that the plain (of drift) on which Elgin stands, and particularly to the east of that town, is underlaid by the cherty or concretionary rock, and it again by the reptiliferous sandstones; but no vestige of reptilian remains has been found within this area.

Appendix IV

Because of Huxley's preoccupation with the notion of persistent types given below is the complete text of a talk he gave at the Royal Institution on Friday 3 June 1859 which was published in summary form with the title 'On the Persistent Types of Animal Life'.

ON THE PERSISTENT TYPES OF ANIMAL LIFE

The successive modifications which the views of physical geologists have undergone since the infancy of their science, with regard to the amount and the nature of the changes which the crust of the globe has suffered, have all tended in one direction, viz., towards the establishment of the belief, that throughout that vast series of ages which was occupied by the deposition of the stratified rocks, and which may be called 'geological time' (to distinguish it from the 'historical time' which followed, and the 'pre-geological time', which preceded it), the intensity and the character of the physical forces which have been in operation, have varied within but narrow limits; so that, even in Silurian or Cambrian times, the aspect of physical nature must have been much what it is now.

This uniformitarian view of telluric conditions, so far as geological time is concerned, is however, perfectly consistent with the notion of a totally different state of things in antecedent epochs, and the strongest advocate of such 'physical uniformity' during the time of which we have a record might, with perfect consistency, hold the so-called 'nebular hypothesis', or any other view involving the conception of a long series of states very different from that which we now know, and whose succession occupied pre-geological time.

The doctrine of physical uniformity and that of physical progression are therefore perfectly consistent, if we regard geological time as having the same relation to pre-geological time as historical time has to it.

The accepted doctrines of palaeontology are by no means in harmony with these tendencies of physical geology. It is generally believed that there is a vast contrast between the ancient and the modern organic worlds – it is incessantly assumed that we are acquainted with the beginning of life, and with the primal manifestation of each of its typical forms: nor does the fact that the discoveries of every year oblige the holders of these views to change their ground, appear sensibly to affect the tenacity of their adhesion.

Without at all denying the considerable positive differences which really exist between the ancient and the modern forms of life, and leaving the negative ones to be met by the other lines of argument, an impartial examination of the facts revealed by palaeontology seems to show that these differences and contrasts have been greatly exaggerated.

Thus, of some two hundred known orders of plants, not one is exclusively fossil. Among animals, there is not a single totally extinct class; and of the orders, at the outside not more than seven per cent. are unrepresented in the existing creation.

Again, certain well marked forms of living beings have existed through enormous epochs, surviving not only the changes of physical conditions, but persisting comparatively unaltered, while other forms of life have appeared and disappeared. Such forms may be termed 'persistent types' of life; and examples of them are abundant enough in both the animal and the vegetable worlds.

Among plants, for instance, ferns, club mosses, and *Coniferae*, some of them apparently generically identical with those now living, are met with as far back as the Carboniferous epoch; the cone of the oolitic *Araucaria* is hardly distinguishable from that of existing species; a species of *Pinus* has been discovered in the Purbecks, and a walnut (*Juglans*) in the cretaceous rocks. All these are types of vegetable structure, abounding at the present day; and surely it is a most remarkable fact to find them persisting with so little change through such vast epochs.

Every subkingdom of animals yields instances of the same kind. The *Globigerina* of the Atlantic foundings is identical with the cretaceous species of the same genus; and the casts of lower Silurian *Foraminifera*, recently described by Ehrenberg, assure us of the very close resemblance between the oldest and the newest forms of many of the *Protozoa*.

Among the *Coelenterata*, the tabulate corals of the Silurian epoch are wonderfully like the millepores of our own sees, as every one may convince himself who compares *Heliolites* with *Heliopora*.

Turning to the *Mollusca*, the genera *Crania, Discina, Lingula,* have persisted from the Silurian epoch to the present day, with so little change, that very competent malacologists are sometimes puzzled to distinguish the ancient from the modern species. *Nautili* have a like range, and the shell of the liassic *Loligo* is similar to that of the 'squid' of our own seas. Among the *Annulosa*, the carboniferous insects are in several cases referable to existing genera, as are the *Arachnida*, the highest group of which, the scorpions, is represented in the coal by a genus differing from its living congeners only in the disposition of its eyes.

The vertebrate subkingdom furnishes many examples of the same kind. The *Ganoidei* and *Elasmobranchii* are known to have persisted from at least the middle of the Palaeozoic epoch to our own times, without exhibiting a greater amount of deviation from the typical characters of these orders, than may be found within their limits at the present day.

Among the *Reptilia*, the highest group, that of the *Crocodilia*, was represented at the beginning of the Mesozoic epoch, if not earlier, by species identical in the essential character of their organization with those now living, and presenting differences only in such points as the form of the articular faces of their vertebrae, in the extent to which the nasal passages are separated from the mouth by bone, and in the proportions of the limbs. Even such imperfect knowledge as we possess of the ancient mammalian fauna leads to the belief that certain of its types, such as that of the *Marsupialia*, have persisted with no greater change through as vast a lapse of time.

It is difficult to comprehend the meaning of such facts as these if we suppose that each species of animal and plant, or each great type organization, was formed and placed upon the surface of the globe at long intervals by a distinct act of creative power; and it is well to recollect that such an assumption is as unsupported by tradition or revelation as it is opposed to the general analogy of Nature.

If, on the other hand, we view 'Persistent Types', in relation to that hypothesis which supposes the species of living beings living at any time to be the result of the gradual modification of pre-existing species – a hypothesis which, though unproven and sadly damaged by some of its supporters, is yet the only one to which physiology lends any countenance – their existence would seem to show, that the amount of modification which living beings have undergone during geological time is but very small in relation to the whole series of changes which they have suffered. In fact, palaeontology and physical geology are in perfect harmony, and coincide in indicating that all we know of the conditions in our world during geological time, is but the last term of a vast and, so far as our present knowledge reaches, unrecorded progression.

Appendix V

On 10 February 1860, Huxley gave a talk at the Royal Institution called 'On Species and Races, and their Origin' which was published in summary form in Volume III of *The Proceedings of the Royal Institution of Great Britain*. He discussed the recently published *On the Origin of Species*, with particular attention to the idea of modification. Was there evidence that a species could be modified by selective breeding? The final section is included here as an Appendix for two reasons. First, it shows Huxley making physiology the ultimate test. Second, it shows him being diverted from a question of science into general propaganda on behalf of science.

It must be admitted that, by selection, a species may be made to give rise experimentally to recessively different modifications; and the next question is: Do causes adequate to exert selection exist in nature? On this point the speaker referred his audience to Mr. Darwin's chapter on the struggle for existence, as affording ample satisfactory proof that such adequate natural causes do exist.

There can be no question that just as man cherishes the varieties he wishes to preserve, and destroys those he does not care about; so nature (even if we consider the physical world as a mere mechanism) must tend to cherish those varieties which are better fitted to work harmoniously with the conditions she offers, and to destroy the rest.

There seems to be no doubt then, that modifications equivalent in extent to the four breeds of pigeons, might be developed from a species by natural causes; and therefore, if it can be shown that these breeds have all the characters which are ever found in species, Mr. Darwin's case would be complete. However, there is as yet no *proof* that, by selection, modifications having the physiological character of species (*i.e.* whose offspring are incapable of propagation, *inter se*) have ever been produced from a common stock.

No doubt the numerous arguments brought forward by Mr. Darwin to weaken the force of this objection are of great weight; no doubt it cannot be proved that all species give rise to hybrids infertile, *inter se*, no doubt (so far as the speaker's private conviction went), a well conducted series of experiments very probably would yield us derivatives from a common stock, whose offering should be infertile, *inter se:* but we must deal with facts as they stand; and at present it must be admitted that Mr. Darwin's theory does not account for all the phenomena exhibited by species; and so far, falls short of being a satisfactory theory.

Nevertheless the speaker expressed his sense of the extremely high value to be attached to Mr. Darwin's hypothesis; and, avowing his own conviction that the following it out must ultimately lead us to the detection of the laws which have governed the origin of species, he concluded his discourse in the following words, which he wishes to be added in full to the very brief preceding account of his view of Mr. Darwin's argument. –

I have endeavoured to lay before you what, as I fancy, are the turning points of a great controversy; to render obvious the mode in which the vast problem of the origin of species must be dealt with; and so far as purely scientific considerations go, I have nothing more to say. But let me beg you still to listen to a last word respecting the unscientific objections which I constantly hear brought forward on the part of the general public, against such doctrines as those we have been discussing. For this is a matter upon which it is of the utmost importance that men of science and the public should come to an understanding. I have heard it said that it is presumptuous for us to attempt to inquire into such matters as these; that they are problems beyond the reach of the human understanding. Do you remember what was the reply of the old philosopher to those who demonstrated to him so clearly the impossibility of motion? 'Solvitur ambulando', said he, and got up and walked. And so I doubt not that one of these days either Mr. Darwin's hypothesis, or some other, will get up and walk, and that vigorously; and so save us the trouble of any further discussion of this objection.

Another, and unfortunately a large class of persons, take fright at the logical consequences of such a doctrine as that put forth by Mr. Darwin. If all species have arisen in this way, say they – Man himself must have done so; and he and all the animated world must have had a common origin. Most assuredly. No question of it.

But I would ask, does this logical necessity add one single difficulty of importance to those which already confront us on all sides whenever we contemplate our relations to the surrounding universe? I think not. Let man's mistaken vanity, his foolish contempt for the material world, impel him to struggle as he will, he strives in vain to break through the ties which hold him to matter and the lower forms of life.

In the face of the demonstrable facts, that the anatomical difference between man and the highest of the *Quadrumana* is less than the difference between the extreme types of the Quadrumanous order; that, in the course of his development, man passes through stages which correspond to, though they are not identical with, those of all the lower animals; that each of us was once a minute and unintelligent particle of yolk-like substance; that our highest faculties are dependent for their exercise upon the presence of a few cubic inches, more or less, of a certain gas in one's blood; in the face of these tremendous and mysterious facts, I say, what matters is whether a new link is or is not added to the mighty chain which indissolubly binds us to the rest of the universe. Of what part of the glorious fabric of the world has man a right to be ashamed – that he is so desirous to disconnect himself from it? But I would rather reply to this strange objection by suggesting another line of thought. I would rather point out that perhaps the very noblest use of science as a discipline is, that now and then she brings us face to face with difficulties like these. Laden with our idols, we follow her blithely – till a parting in the roads appears, and she turns, and with a stern face asks us whether we are men enough to cast them aside, and follow her up the steep? Men of science are such by virtue of having answered her with a hearty and unreserved, Yea; by virtue of having made their election to follow science whithersoever she leads, and whatsoever lions be in the path. Their duty is clear enough.

And, in my apprehension, that of the public is not doubtful, I have said that the man of science is the sworn interpreter of nature in the high court of reason. But of what avail is his honest speech if ignorance is the assessor of the judge, and prejudice foreman of the jury? I hardly know of a great physical truth, whose universal reception has not been preceded by an epoch in which most estimable persons have maintained that the phenomena investigated were directly dependent on the Divine Will, and that the attempt to investigate them was not only futile, but blasphemous. And there is a wonderful tenacity of life about this sort of opposition to physical science. Crushed and maimed in every battle, it yet seems never to be slain; and after a hundred defeats it is at this day as rampant, though happily not so mischievous, as in the time Galileo.

But to those whose life is spent, to use Newton's noble words, in picking up here a pebble and there a pebble on the shores of the great ocean of truth – who watch, day by day, the slow but sure advance of that mighty tide, bearing on its bosom the thousand treasures wherewith man enobles and beautifies his life – it would be laughable, if it were not so sad, to see the little Canutes of the hour enthroned in solemn state, bidding that great wave to stay, and threatening to check its beneficent progress. The wave rises and they fly; but unlike the brave old Dane, they learn no lesson of humility; the throne is pitched at what seems a safe distance, and the folly is repeated.

Surely it is the duty of the public to discourage everything of this kind, to discredit these foolish meddlers who think they do the

Almighty a service by preventing a thorough study of his works.

The Origin of the Species is not the first, and it will not be the last of the great questions born of science, which will demand settlement from this generation. The general mind is seething strangely, and to those who watch the signs of the times, it seems plain that this nineteenth century will see revolutions of thought and practice as great as those which the sixteenth witnessed. Through what trials and sore contests the civilized world will have to pass in the course of this new reformation, who can tell?

But I verily believe that come what will, the part which England may play in the battle is a grand and a noble one. She may prove to the world that for one people, at any rate, despotism and demagogy are not the necessary alternatives of government; that freedom and order are not incompatible; that reverence is the handmaid of knowledge; that free discussion is the life of truth, and of true unity in a nation.

Will England play this part? That depends upon how you, the public, deal with science. Cherish her, venerate her, follow her methods faithfully and implicitly in their application to all branches of human thought; and the future of this people will be greater than the past.

Listen to those who would silence and crush her, and I fear our children will see the glory of England vanishing like Arthur in the mist; they will cry too late the woeful cry of Guinevere:

> 'It was my duty to have loved the highest
> It surely was my profit, had I known;
> It would have been my pleasure had I seen.'

Appendix VI

A SELECT LIST OF WORKS PERTINENT TO THE PRESENT STUDY

Agassiz, Louis, *Recherches sur les poissons fossiles (Neuchâtel,* 1833–43).

Agassiz, Louis, *An Essay on Classification*, London, 1859.

Avebury, John Lubbock, Lord, 'Huxley's Life and Work', *Nature*, 63 (1900): 92–96, 116–19

Benton, M. J., *The Reign of the Reptiles* (London: Kingfisher Books, 1990).

Benton, M. J., 'The Triassic Reptile *Hyperodapedon from Elgin: functional morphology and relationships*', *Philosophical Transactions*, B, *302*, 605–720.

Boulanger, G. A. 'On Reptilian Remains from the Trias of Elgin', *Philosophical Transactions*, B, *196*, 175.

Brickenden, L. B., 'Notice of the Discovery of Reptilian Foot-tracks and Remains in the Old Red or Devonian Strata of Moray', *Journal of the Geological Society*, 8 (1852): 100–05.

Cuvier, G., *Recherches sur les ossements fossiles.*

Falconer, H. 'On Professor Huxley's Attempted Refutation of Cuvier's Laws of Correlation in the Reconstruction of Extinct Vertebrate Form', *Annals and Magazine of Natural History*, 17 (1856): 476–93.

Haeckel, E., *Thomas Henry Huxley, Nature*, 9 (1871): 257–58.

Haeckel, E., *Generelle Morphologie des Organismen* (Berlin, 1866).

Jensen, J. V. 'The X-Club: Fraternity of Victorian Scientists', *The British Journal from the History of Science*, 5 (1970): 63–72.

Marsh, O. C. 'Thomas Henry Huxley', *American Journal of Science*, 50 (1895): 177–83.

Marsh, O. C. 'Notice of a New and Diminutive Species of Fossil Horse (*Equus parvulus*) of Nebraska', *American Journal of Science and Art*, 46 (1868): 374–5.

Marsh, O. C. 'Fossil Horses in America', *The American Naturalist*, 8 (1874): 288–94.

Murchison, R. I. 'On the Sandstones of Morayshire (Elgin, etc.) Containing Reptilian Remains; and on Their Relations to the Old Red Sandstone of that Country', *Quarterly Journal of the Geological Society*, 15 (1858): 419–39.

Newton, E. T., *On the Reptilia of the British Trias*, British Association (Nottingham), *63*, 752.

Newton, E. T., 'Reptiles from the Elgin Sandstone – description of two new genera', *Philosophical Transactions*, B, *185*, 573.

Owen, R., *On the classification and Geographical Distribution of the Mammalia* (London, 1859).

Paradi, J. G., *T. H. Huxley: Man's Place in Nature* (Lincoln: University of Nebraska Press, 1978).

Pingree, J., *T. H. Huxley: A List of his Scientific Papers.*

Pingree, J., *Thomas Henry Huxley: List of his Correspondence with Miss Henrietta Heathorn, later Mrs. Huxley, 1847–54.*

Porter, Roy, *The Making of Geology Earth Science in Britain 1660–1815* (Cambridge: Cambridge University Press, 1977).

Romer, A. S., *Osteology of the Reptiles* (University of Chicago Press, 1956).
Rudwick, M. J. S., *The Great Devonian Controversy* (Chicago: University of Chicago Press, 1985).
Secord, James, *Controversy in Victorian Geology: the Cambrian-Silurian Dispute* (Princeton: Princeton University Press, 1986).
Walker, A. D., *Triassic Reptiles from the Elgin Area: Stagonolepis, Dasygnathus and their Allies, Philosophical Transactions*, B, *709*, 103–204.
Walker, A. D., *Triassic Reptiles from the Elgin Area: Ornithosuchus and the Origin of Carnosaurs, Philosophical Transactions*, B, *744*, 53–134.

Index